W9-BOB-865

An OPUS book

The Problems of Chemistry

OPUS General Editors

OPUS books provide concise, original, and authoritative introductions to a wide range of subjects in the humanities and sciences. They are written by experts for the general reader as well as for students.

The Problems of Science

This group of OPUS books describes the current state of key scientific subjects, with special emphasis on the questions now at the forefront of research.

The Problems of Physics A. J. Leggett
The Problems of Biology John Maynard Smith
The Problems of Chemistry W. Graham Richards
The Problems of Evolution Mark Ridley
The Problems of Mathematics Ian Stewart

The Problems of Chemistry

W. GRAHAM RICHARDS

OXFORD UNIVERSITY PRESS
New York Oxford

Oxford University Press

Oxford New York Toronto
Delhi Bombay Calcutta Madras Karachi
Petaling Jaya Singapore Hong Kong Tokyo
Nairobi Dar es Salaam Cape Town
Melbourne Auckland

and associated companies in
Berlin Ibadan

Published by Oxford University Press, Inc.,
200 Madison Avenue, New York, New York 10016

First published in 1986 as an Oxford University Press paperback and simultaneously in a hardback edition, Walton Street, Oxford OX2 6DP

British Library Cataloguing in Publication Data
Richards, W. G.
The problems of chemistry.—(OPUS)
1. Chemistry
I. Title II. Series
540 QD31.2
ISBN 0–19–219191–8
ISBN 0–19–289172–3 Pbk

Library of Congress Cataloging in Publication Data
Richards, W. G. (William Graham)
The problems of chemistry
(An OPUS book)
Bibliography: p.
Includes index.
1. Chemistry. I. Title. II. Series: OPUS.
QD31.2.R53 1986 540 85–13715
ISBN 0–19–219191–8 (U.S.)
ISBN 0–19–289172–3 (U.S.: pbk.)

2 4 6 8 10 9 7 5 3 1

Printed in the United States of America

Preface

Chemistry had a beginning. It can be dated fairly precisely as a few hundred thousand years after the 'big bang', which is time zero for the universe as we comprehend it. By then, long before galaxies or solid bodies were in existence, molecules had formed. The formation of molecules and their subsequent reactions to form new molecules, then solids, and ultimately our world, are the stuff of chemistry. Chemistry as an academic discipline is predominantly the study of phenomena at the level of the molecule.

The molecule is the smallest unit of matter which is stable under the conditions prevailing in our world, and therefore the study of phenomena at the molecular level is inevitably wide-ranging. The basic unit of almost everything we see, touch, feel, or indeed are is molecular. Thus chemistry encompasses molecular biology, the nature of materials, and the properties of gases both in the atmosphere and in interstellar space.

That molecules are the building blocks of matter did not become apparent until the nineteenth century, because most substances are complicated mixtures of already complex molecules and were beyond the analytical understanding of early chemists. Once scientists started defining the molecular nature of substances, however, progress was explosive. By the present time, some 20 million separate and distinguishable molecules have been characterized. Theory has advanced to the point where the structures of many molecules can be predicted from a knowledge of their constituent building blocks—atoms. At the same time, the results of mixing molecules to give reactions which create new molecules are still largely deduced from experiment, and even if the products of such reactions can sometimes be predicted, the rate (or speed) of the reactions remains a matter of speculation.

By comparison with its flanking subjects, physics and biology, chemistry often seems unspectacular. As a discipline it does not enjoy the Press or television coverage given to quarks and particle physics, to black holes and astronomy, or to advances in biology and medicine. Despite this, chemistry, of all academic disciplines, has had the most far-reaching impact on social habits and values. The atomic bomb may have been spectacular, but the lives of people have been influenced far more by less dramatic innovations such as detergents, artificial fibres, plastics, synthetic dyes, antibiotics, and the contraceptive pill: all the fruits of chemistry.

The author is happy to acknowledge the generous editorial help of Hugh Oliver.

Contents

List of figures

The publishers wish to express their gratitude to Thomas Nelson and Sons for their permission to reproduce the illustration on p. 13.

1

How it all started: molecules in interstellar space

In the beginning, according to the best available theory, there was a huge explosion. The 'big bang', as physicists and cosmologists call it, is generally agreed to have been the unique moment when our universe began. It can be dated to a time some 13,000 million years ago when, with all matter/energy confined in one spot, there was an explosion which sent every particle of matter rushing apart from every other particle. This process continues to our own day as galaxies recede from each other: hence the description 'expanding universe'.

The formation of atoms

Physicists' knowledge of the early universe immediately after the big bang now extends back to less than one-hundredth of a second after the primary event. At this stage, the temperature of the universe was hotter than the centre of the sun and its density more than a billion times denser than water. Since that time, the universe has cooled and its density has diminished, at first rapidly and later more slowly. As the cooling has proceeded, the fundamental constituents of matter have gradually accreted to give more complex, heavier structures—just as, on a simpler scale, steam cools to give first water, then ice.

In the early minutes of the universe, the only matter existing was in the form of the fundamental particles which are the building blocks of atomic nuclei (the cores of atoms). In our time, to study these particles we have to expend massive amounts of energy and do collision experiments to smash up atoms. Such experiments in particle physics are performed with gigantic and expensive accelerators, like those at the CERN (European Centre for Nuclear Research) laboratories in Geneva or the Stanford Linear Accelerator in California.

Some three minutes after the big bang, the inverse of the process we use in atomic bombs or for nuclear power generation occurred: the creation of matter from energy.

In nuclear power-stations, the energy holding together the fundamental particles which make up the nuclei of atoms is released as heat and is used to generate steam to drive turbines: energy is created from matter following Einstein's famous equation, $E = mc^2$—where the energy, E, is related to the mass, m, by the large constant, c^2, the square of the velocity of light. So fast is the speed with which light travels (186,000 miles per second) that only small amounts of matter in the form of uranium or plutonium are needed to create vast quantities of energy. In the early universe, the equation worked in reverse, and energy in vast amounts was used to produce relatively small amounts of matter, some of it in the form of positively charged atomic nuclei and other building blocks of matter like electrons.

The positive electrical charges on these atomic nuclei attracted the very much smaller, negatively charged electrons—just as the north and south poles of magnets attract each other. Working against the attractive forces were the sheer chaos and random motions of the heat-activated particles—like a set of magnets shaken so violently that the attractions are not strong enough to overcome the buffeting as they bang into each other. After a few hundreds of thousands of years, however, the universe cooled sufficiently for the attractive forces to dominate the random chaos. Negatively charged electrons joined with positively charged nuclei to form neutral atoms.

At the same time, different types of atom arose from the fusion of nuclei, resulting in about a hundred distinct chemical elements. (To be precise, there are 92 naturally occurring elements, uranium being the heaviest; and since 1940, about a dozen, heavier, man-made elements have been synthesized.) Each element is distinguishable by the number of positive charges on the nucleus, known as the atomic number. In the electrically neutral atom, the nucleus is surrounded by a corresponding number of negative electrons. Neutral atoms form the basic building units of molecules. If the atom gains or loses an electron it will have a resultant negative or positive charge and is often called an ion.

Figure 1 shows the structures of some of the simpler elements, and Figure 2 part of the periodic table, in which is summarized the structures and many of the properties of the different elements. The periodic table was first devised in 1886 by the Russian chemist Mendeleyev, who was sent to study in France and Germany. Mendeleyev sought to classify the chemical elements according to the weights of their respective atoms (atomic numbers being unknown at that time). At a conference in Karlsruhe in 1860, many of the outstanding questions about atomic weights were settled, providing Mendeleyev with the data

Hydrogen

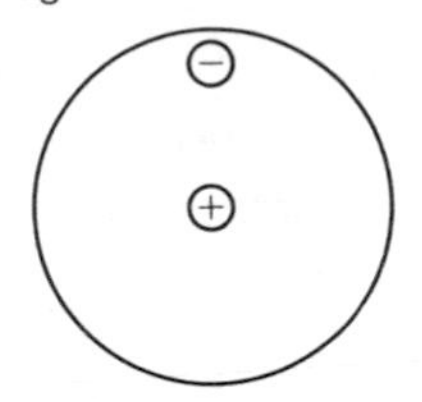

Symbol H, atomic number 1
Nucleus has one proton ⊕
One electron ⊖ orbits the nucleus

Helium

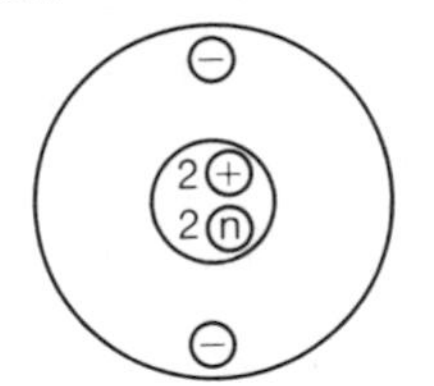

Symbol He, atomic number 2
Nucleus has two protons ⊕ and two neutrons ⓝ
Two electrons ⊖ orbit the nucleus

Carbon

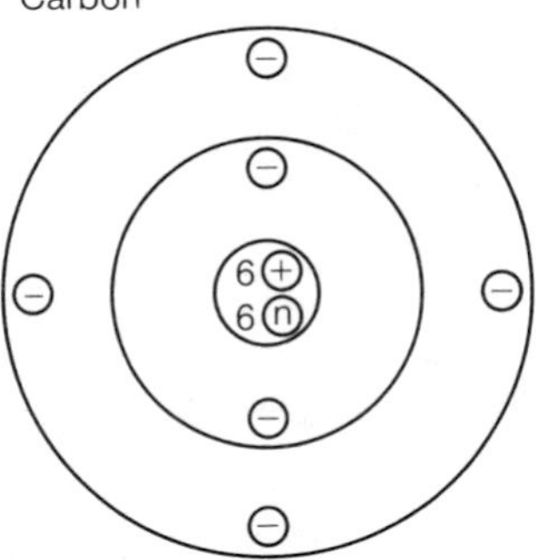

Symbol C, atomic number 6
Nucleus has six protons ⊕ and six neutrons ⓝ
Six electrons ⊖ orbit the nucleus in two shells.

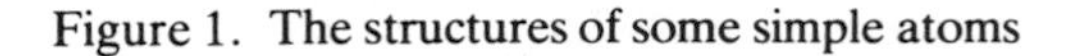

Figure 1. The structures of some simple atoms

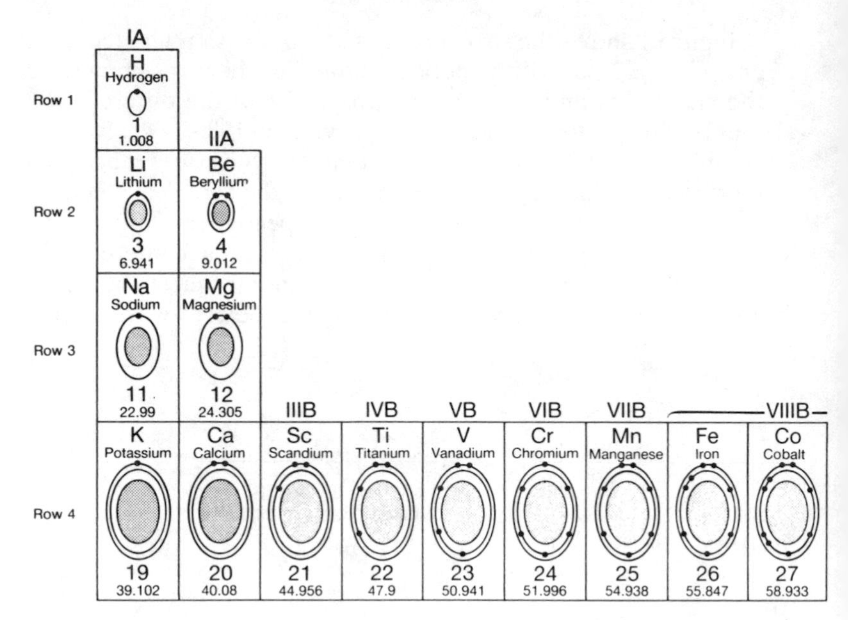
IA
IIA
IIIB
IVB
VB
VIB
VIIB
VIIIB
Row 1
Row 2
Row 3
Row 4
H
Hydrogen
1
1.008
Li
Lithium
3
6.941
Be
Beryllium
4
9.012
Na
Sodium
11
22.99
Mg
Magnesium
12
24.305
K
Potassium
19
39.102
Ca
Calcium
20
40.08
Sc
Scandium
21
44.956
Ti
Titanium
22
47.9
V
Vanadium
23
50.941
Cr
Chromium
24
51.996
Mn
Manganese
25
54.938
Fe
Iron
26
55.847
Co
Cobalt
27
58.933

Figure 2

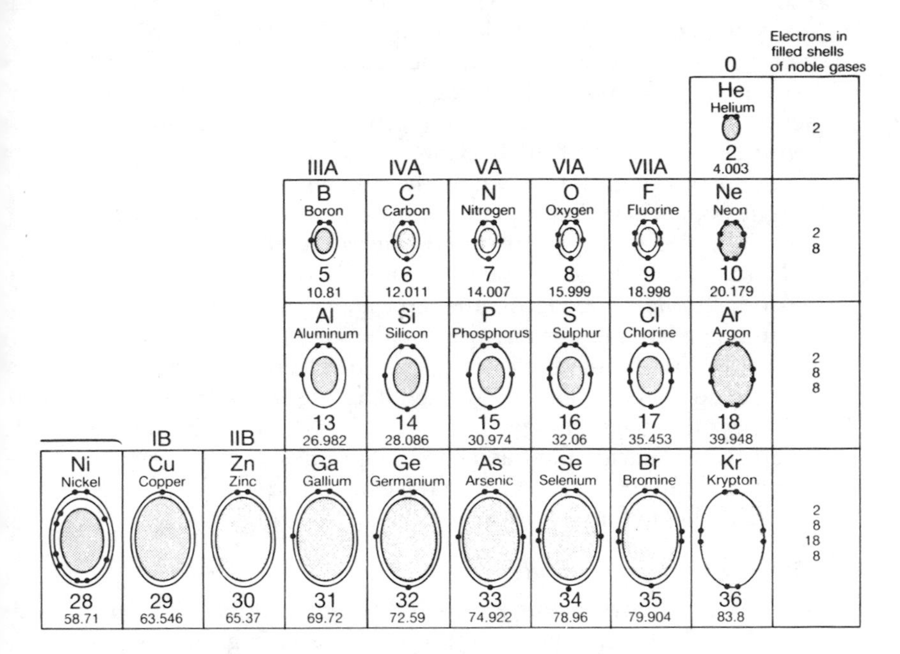

Figure 2. Part of the periodic table of elements

The progressive buildup of atoms is shown by the filling of shells with electrons (dots). The first-row shell is filled with 2 electrons; the second-row shell with 8 electrons; and the third-row shell also with 8 electrons. The fourth row adds 18 electrons in two subgroups which accept up to 10 and 8 electrons respectively. The atomic weights of the elements are the small numbers under the atomic numbers. They are all based on an assignment of exactly 12 as the atomic weight of the carbon-12 isotope.

upon which to make his classification. Arranged in order of increasing atomic number, the chemical elements exhibit some properties in common down the table and, to a lesser extent, across the table. For example, the properties of magnesium are quite similar to those of beryllium and calcium above it and below it in the group known as the alkaline earth metals, and magnesium also shares some properties with sodium and aluminium on either side of it in the horizontal period.

Modern knowledge has revealed that the order inherent in the periodic table depends on the arrangement of the electrons around the atomic nucleus or, in other words, on the way in which the atomic number is made up. The electrons occupy what are called 'shells', and each shell has a specific number of electrons that it can contain before it becomes 'full'. The number of electrons outside the filled shells is primarily responsible for the chemical properties of an element, including the types of molecule it will form.

The formation of molecules

For the most part, atoms are joined to other atoms. Although they are electrically neutral, atoms have a strong tendency to combine so that their electrons can form certain stable groupings. A situation with two electrons orbiting an atomic nucleus is stable in a way in which a single electron orbiting a nucleus is not. Thus hydrogen, with its single negative electron orbiting the nuclear charge of one positive unit, is not stable as a single atom. By combining, two hydrogen atoms form the more stable hydrogen molecule, in which the two electrons orbit both of the two nuclei.

The arrangements of electrons which have the greatest stability are revealed by looking at one particular group of elements in the periodic table—the rare or inert gases, helium, neon, argon, krypton, xenon, and radon. The atoms of these gases do not normally combine with other atoms. Hence the arrangement of electrons in their shells must have special stability. The stable totals for the electron shells, as revealed by the electron arrangements in these rare gases, are 2, 8, 8, 18, 18, and 32. The precise significance of these specially stable groupings is now readily understandable from quantum mechanics, which can predict how many electrons confer special stability on particular shells.

In elements other than the rare gases, similar shell structures of electrons are found, but the outermost shell is incomplete and can achieve the special stability only by gaining or losing electrons (which will leave the atom negatively or positively charged) or by sharing electrons (as in the hydrogen molecule). The tendency of atoms to combine so as to produce stable electronic arrangements leads to the formation of molecules. Among the different chemical elements, there are millions of possible combinations. Simple examples of molecules are water (H_2O), ammonia (NH_3), and alcohol (C_2H_5OH). These formulae indicate which atoms of a particular type are present in molecules of a pure substance and how many there are of each type.

Molecules, as mentioned at the beginning, are the building blocks of all the substances we see and experience in everyday life.

Molecular models

The molecule is the building block which helps to explain the widest range of phenomena in medicine and biology as well as in chemistry. The simplest and most convenient way of representing the molecular idea is through models. Atoms can be thought of as spheres whose sizes, known from experiment, depend on the number of orbiting electrons (equal to the positive nuclear charge) they contain and the shell structure of these electrons. Indeed, atoms are conventionally represented by coloured spheres of appropriate relative sizes. Each sphere is a scaled-up representation of an atom, the size increased by approximately one hundred million times (Figure 3 gives some simple examples).

The models may also show how the atoms combine to form the stable electron pairings of the rare gases. For example, hydrogen needs one more electron whereas carbon needs four more electrons for a stable grouping to be obtained. And it is here worth noting that chemical elements in the same vertical group in the periodic table have similar electronic arrangements (especially in terms of the number of electrons needed to fill their outer shells), which explains why their chemical properties are similar.

Where electrons are shared between atoms, chemists speak of 'bonds' being formed. These electron-shared bonds (covalent bonds as they are sometimes called) are strong, and at normal

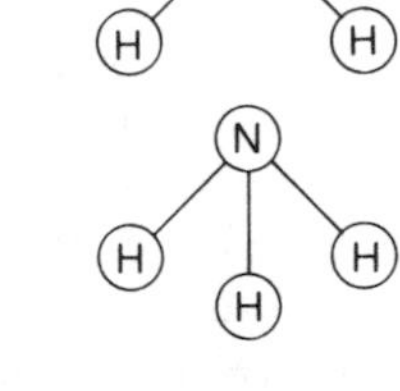

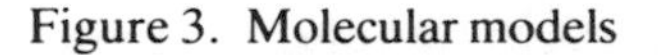

Figure 3. Molecular models

temperatures the molecules formed by this bonding of atoms are quite stable. Thus, although in heating water we may separate the molecules from the weak forces which hold them together as a liquid, we do not disintegrate the individual water molecules (H_2O) into hydrogen and oxygen unless we use very large amounts of energy.

Interstellar molecules

Some molecules were formed in the early years of the universe by atoms combining to give systems of shared electrons which became more stable as the temperature fell. As the fireball expanded and cooled, gravity caused some of the atomic and molecular matter to aggregate and to form clusters of galaxies and individual galaxies. Within each galaxy, clouds of atomic and molecular particles condensed to form dust particles, leading to further aggregation, and, ultimately, to the formation of stars like our own sun. In the hot interior of stars, nuclear reactions give rise to the hundred or so different sorts of atom (i.e. chemical elements), which, as we have seen, combine to give millions of different types of molecule.

We know there are molecules in interstellar space because we can detect them using radioastronomy. Since 1964, when the simple molecular species OH was discovered, the list has grown rapidly to include over fifty examples, ranging from two-atom molecules like carbon monoxide to molecules with a dozen or more atoms joined in a line.

Radioastronomers detect molecules by picking up the energy they emit as their rotational energy drops. Alternatively, light from a star may be absorbed by an intervening cloud of molecules, and this absorption of energy is then detected with a radio telescope. The problems for the chemist are firstly observational—deciding what molecules are present—and then more abstract—theorizing about reactions which would explain the known concentrations of the molecules.

The most common chemical element between the stars is hydrogen. Even though the density there is very low indeed, there are many molecules because of the vastness of space. In 1974 it was estimated that in the cloud in Sagittarius, there are as many molecules of alcohol as there are in 100,000,000,000,000,000,000,000 bottles of whisky!

The currently surprising feature of the molecules found in the interstellar regions is that they include some of the fundamental building blocks of life and also big reactive molecules from which the chemist could synthesize quite complicated molecular structures, almost as complex as the simplest living things.

At its most basic level, 'life' consists of molecules which can reproduce themselves in some fashion. Viruses, although not living, cause replicas of themselves to be made by subverting the synthetic machinery of the cells they infect by introducing molecules of their own type. Most scientists are of the persuasion that life began on earth in its early history; but the possibility that this significant step began extra-terrestrially and that life was brought to earth, perhaps on a comet, is espoused by some eminent scientists, including the British astronomer Fred Hoyle.

Life does exist on earth, either because it was brought from outside or, as seems more likely and as we shall investigate later, because replicating or reproducing molecules were formed on the surface of the young earth in what has been termed the 'primordial soup'. By natural selection at the molecular level, this process led to the life forms we know on earth, all made from molecules.

2

The nature of chemistry

Since chemistry is the science of molecules, it could not make much progress until it was possible both to count and to weigh individual atoms and molecules. When weighing anything, we always compare the weight of whatever it is we are interested in against some standard. Talking of our own weight, we say we weigh so many pounds or so many kilograms, referring our own body weight to the defined standard of the pound or the kilogram. At the atomic level, were it possible to measure an atom's weight (and it is), an obvious standard would be the weight of the nucleus of the hydrogen atom, hydrogen having the lightest and simplest atom of all the chemical elements.

Atomic nuclei

The nucleus of the hydrogen atom with its single positive charge is called the proton, and round it orbits the single, much lighter, negative electron. We can understand the stability of this system with a model, devised by Ernest Rutherford and Nils Bohr, in which the electron orbits the proton like the Moon orbits the Earth. The stability arises from the balancing of the attraction of the positive (nuclear) and negative (electronic) charges with the centrifugal effect of the orbiting mass (the electron)—a centrifugal effect similar to that when one swings a weight on a string round one's head. The force attracting the Moon to the Earth is, of course, gravitational not electrical, but the nature of the centrifugal force is the same.

The proton is so much heavier than the electron (almost 2,000 times) that if we take the mass of the proton as 1 unit of mass, the mass of a hydrogen atom (proton plus electron) will still be roughly one unit.

Another particle found in atomic nuclei is the neutron, which has the same mass as the proton but no electrical charge. It is the charge on the atomic nucleus (equal to the number of orbiting electrons) which determines the type of chemical element of which the atom is the basic unit. Thus if we add a neutron to the hydrogen nucleus (i.e. the proton), we would have an atom of weight two but with a single positive charge; still an atom of hydrogen. Atoms of the same chemical element that differ in weight are called *isotopes*. Sometimes isotopes are given specific chemical names. The heavier hydrogen isotope, with a neutron added to the proton, is known as heavy hydrogen or deuterium. In heavy water, the two normal hydrogen atoms in an ordinary water molecule (H_2O) are replaced by the heavier hydrogen isotope; the heavy water formula is written either as D_2O or, more formally, as ^2H_2O, with the superscript indicating the rough mass of the nucleus. The atomic properties of heavy water resemble those of ordinary water but there are important differences in physical properties.

Many of the atoms of the periodic table of elements exist in several isotopic forms. Occasionally their nuclei may be unstable, decaying into other types of nucleus by emitting particles and energy. Isotopes which do this are radioactive. The best-known example is the uranium atom, which has 92 protons and over 100 neutrons. One of uranium's isotopes, with a mass of 238, (92 protons, 146 neutrons), is quite stable; another isotope, with a mass of 235 (with three less neutrons) is radioactive. The naturally occurring element is a mixture of the two isotopes so that, in order to use the radioactive form ^{235}U for power generation or the manufacture of nuclear weapons, the two isotopes have first to be separated. Separating the isotopes is a problem which has taxed, and continues to tax, the minds of many chemists.

Atomic and molecular weights

The oxygen atom has eight protons and eight neutrons, and hence an approximate weight of 16, each proton and each neutron having an approximate mass of 1 unit. If two hydrogen atoms combine with one oxygen atom to give one molecule of water (H_2O), then the water molecule must weigh approximately 18 units (assuming, which we do, that no mass is created or destroyed). Likewise, 20 atoms of hydrogen (total weight 20 units) will combine with 10 atoms of oxygen (total weight 160 units) to give 10 molecules of water (total weight 180 units). We could scale this reaction up until we were combining many millions of atoms, as in a real experiment. The ratio of weights would always remain the same: 2 tons of hydrogen atoms would combine with 16 tons of oxygen atoms to give 18 tons of water.

Chemists generally work in grams as a convenient measure of weight. The atomic weight of an element expressed in grams will contain a vast number of atoms but it will be the same number of atoms for all elements: one gram of hydrogen will contain the same number of atoms as 16 grams of oxygen because each individual oxygen atom is 16 times as heavy as each hydrogen. This number of atoms in the atomic weight of an element expressed in grams is known as Avogadro's number, after the Italian chemist of that name, and is roughly equal to 600,000,000,000,000,000,000,000 (or 6×10^{23}) atoms, which gives some idea how small and how light an individual atom must be.

The molecular weight of a substance is the sum of the weights of the atoms which comprise the molecule—in the case of water, 18. Thus there will be Avogadro's number of water molecules in 18 grams of water. Chemists describe a substance's molecular weight expressed in grams as a *mole* of the substance. Moles are convenient quantities to work with experimentally; furthermore the weights reflect the actual numbers of individual atoms or molecules, and we can therefore interpret what is going on at the atomic or molecular level. It is easy to weigh 18 grams of water, and this quantity will contain the same number of molecules as 16 grams of oxygen (O_2).

The weights of individual atoms and molecules were first measured in the early nineteenth century by John Dalton, the son of a Cumbrian weaver. He described his atomic theory in a lecture at the Royal Institution in 1803. The key points of his theory were that matter consists of atoms which cannot be created or destroyed (the conversion of matter to energy was then unknown); that all atoms of the same element are identical (again, isotopes were then unknown) but different elements have different atoms; that chemical reactions take place by the rearrangement of atoms; and that compounds or molecules are formed as a result of these rearrangements. Although Dalton devised methods of comparing the weights of atoms, his experimental equipment was not very accurate. Nowadays a mass spectrometer may be used to measure atomic and molecular weights.

In the mass spectrometer (Figure 4(a)), individual atoms or molecules are induced to lose an electron by being bombarded with a stream of electrons. This leaves the atom or molecule with one fewer electron than the number of protons and, consequently,

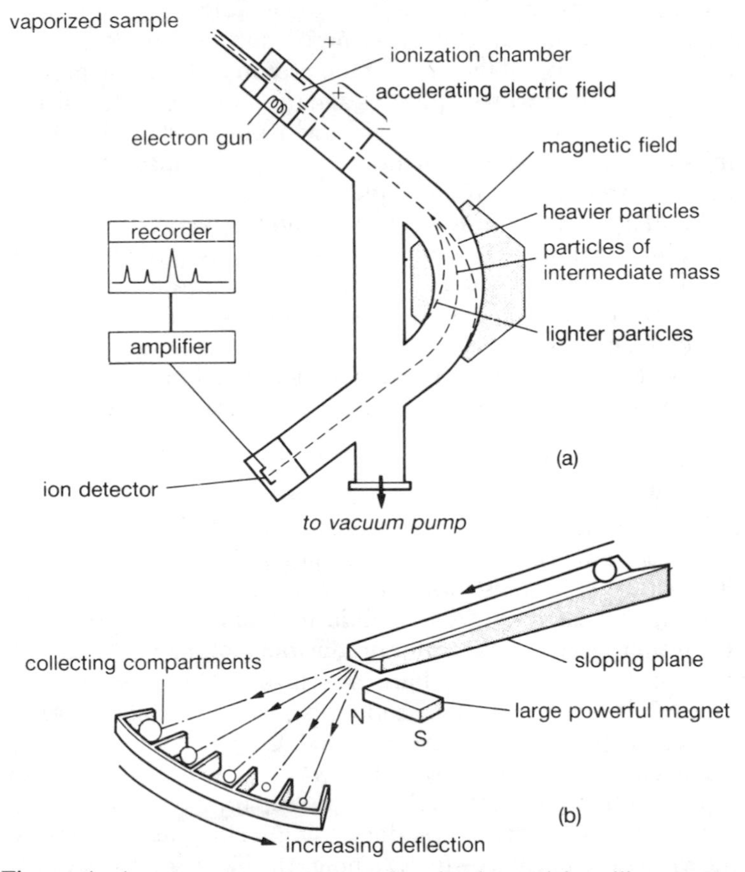

Figure 4. A mass spectrometer and a simple model to illustrate its working

From G. C. Hill and J. S. Holman, *Chemistry in Context* (2nd edn.), Thomas Nelson and Sons (1983).

a net positive charge. An atom or molecule that has a positive or negative charge is called an *ion*. Ions can be deflected by an electric field—the space between two plates of metal, one positively charged and the other negatively charged. This parallels the deflection of iron balls by a magnetic field and is illustrated in Figure 4(b). The amount of deflection depends on the weight of the atom or molecule, and so by measuring the deflection, the weight of the atom or molecule can be determined.

For purely practical reasons, the atomic weight scale (or, more precisely, the scale of relative atomic or molecular mass) takes as its standard unit not the mass of the proton as 1 but the mass of the carbon atom, with six protons and six neutrons, as 12. On this scale, the hydrogen atom weighs not 1 but 1.008 atomic mass units (amu) and the oxygen atom 25.999 amu. Accurate atomic masses differ slightly from whole numbers, stemming from the fact that mass (m) and energy (E) are interconvertible, following Einstein's equation $E = mc^2$. The massive energies implied by small weight discrepancies are the energies used to hold nuclei together. These energies manifest themselves only when nuclei are broken down in nuclear fission or combined to give new nuclei in nuclear fusion. These fission and fusion processes are the bases for the two types of method for extracting energy from atoms.

Pure substances

Pure substances contain atoms or molecules of just one single type. Pure gold contains just gold atoms and pure water contains just water molecules, whereas sea-water contains dissolved substances such as salt. A good indication of a substance's purity is whether it has a clear and reproducible melting-(or freezing-) point and boiling-point. Pure water under a pressure of 1 atmosphere freezes at 0 °C and boils at 100 °C, but salt-water freezes below 0 °C and boils above 100 °C.

The chemist, when studying a substance, will usually want to know whether it is pure, and if not, to separate the mixture into its components. This can be done using techniques which were known to the ancients—for example, distilling solutions to get pure water (by boiling the solution and condensing the steam) or distilling fermentation mixtures to extract alcohol. Other commonly used tricks include sieves with pores of molecular size or one of the many variants of the technique known as chromatography. In chromatography, the separation of a mixture of molecules depends on variation among different types of molecule with respect to sticking to a surface or to another type of molecule in solution. A simple example of chromatography is the separation of the mixture of dyes in blue-black ink by letting the ink seep up a piece of white blotting-paper: the different components rise to different heights and are thus separated. Similarly, when a mixture of gases is forced through a plug of charcoal, the different components may vary in the speed with which they pass through, so that once more a separation can be achieved. Once a chemist has what is thought to be a single pure substance, the next problem is to determine the nature of that substance. This is a task for an analytical chemist, who first determines the nature of the components in a mixture, and then the relative amounts of each.

Analytical chemists

There are millions of different types of molecule, leading to increasingly fine subdivisions of the specialists within chemistry. Unfashionable but extremely important are the analytical chemists. Qualitatively, they answer the question: what molecules are in a mixture? Quantitatively, they answer the question: how much of each particular molecule is there?

Analytical chemists employ a variety of techniques, but most often they make use of the physical properties of known molecules to study the unknown mixture. The absorption of light is a widely used property that enables different molecules in a mixture to be identified. This is an extension of the obvious idea that different substances may be different colours. A complicated molecule may be converted to simpler molecules, the quantities of which tell the chemist how much of a particular element was in the original molecule. For instance, burning a complex hydrogen-containing molecule may convert all the hydrogen atoms into water molecules. Measuring how many water molecules result from the combustion will then indicate how many hydrogen atoms there must have been in the original molecules.

Analytical chemists not only provide a service for other chemists but they also are widely employed in industry, testing for the purity of products and investigating the results of synthetic chemists who have been trying to make new molecules. In addition, they are the major source of scientific forensic evidence: was there arsenic in the stomach of the deceased, or how much alcohol was there in the urine of the drunken driver? The problem facing analytical chemists is to devise ever more accurate and sophisticated techniques to detect smaller and smaller quantities of substances.

Synthetic chemists

As is obvious from the name, the business of synthetic chemists is to make molecules. They are divided into two broad categories: organic chemists and inorganic chemists.

Originally, organic substances were defined as those found in living systems; they were thought to be made by God and not reproducible by Man. An important breakthrough occurred in 1828 when the first organic substance, urea, was synthesized by Wöhler, leading to the modern definition of an organic substance as one which contains one or more carbon atoms. The relation to the older definition is extremely close. Carbon atoms can form chains and rings of atoms, thereby producing a bewildering range of molecules (more than ten million are known and characterized), and nature has used these substances to provide all the building blocks of living systems. The individual organic molecules can usually be burned and are thus not stable at very high temperatures. Organic chemists extract and study the molecules found in nature and also create their own novel variants. New organic molecules may be of interest because of the light which they shed on the principles governing the reactions and stabilities of molecules, or because they have useful applications, perhaps as drugs or dyes or pesticides.

Inorganic chemistry is concerned with all the molecules formed by the hundred or so chemical elements other than carbon. Currently, a major area of interest concerns the properties of inorganic molecules in the solid state. The synthetic inorganic chemist is attempting to create new substances with electrical properties which will be of use in electronic devices as conductors, rectifiers, and memory elements. These will find applications in computers and, perhaps, in systems designed to convert the light from the sun directly into energy—by, for example, splitting water into hydrogen and oxygen. Many experts suggest that hydrogen may be the fuel of the future; heat can be obtained by the reverse process, burning the hydrogen with oxygen.

There are some borderline cases when it comes to whether a substance belongs to organic or inorganic chemistry, notably the organometallics. These substances contain organic groups of atoms, in particular combinations of carbon and hydrogen, and also metal atoms, which are the province of inorganic chemistry. The organometallics have great commercial importance in that they are used in the manufacture of plastics and are of theoretical interest in that they resemble the enzyme molecules which promote the chemical reactions in living systems.

Physical chemists

The principles which govern the stability and reactions of molecules ultimately depend on the fundamental laws of nature—that is, on physics. The physical chemist tries to determine the principles and to come up with both explanations and predictions. Applications of the quantum theory (which provides mathematical equations from which the energy of a molecule may be calculated), to the structure of matter—especially in the 1930s by such people as Schrödinger, Heisenberg, Pauli, and Dirac—have been notably successful in predicting molecular structure, or how the atoms are disposed in space in a stable molecule. It is possible, for example, to calculate theoretically the angle between the O–H bonds in water, and the calculation is in close agreement with the experimental result (about 104°). Less success has been achieved in predicting the rates of chemical reactions.

Physical chemists in the laboratory spend a lot of time devising experiments in which they can study very simple reactions and fundamental processes in order to give the theoretical chemist hard data upon which to test a theory or prediction. A major impetus has been given to physical chemists by the advent of the laser. With this tool, it is possible to put sharply defined amounts of energy into a molecule and then study the effects and ramifications.

3

The earth's atmosphere—gases

Most of the substances which one encounters in everyday life are fairly complicated, involving many thousands of atoms. It is for this reason that the alchemists of old and even early scientists did not make much headway with chemistry. The single obvious exception to this complexity is the air about us and, indeed, gases in general. That the alchemists had little understanding of air is because they lacked a method for separating and collecting its component gases.

Gases in the atmosphere

The air we breathe is made up of about 80 per cent nitrogen and 20 per cent oxygen—the diatomic or two-atom molecules N_2 and O_2. With N_2, all the electrons are paired, and none are left to form bonds with other substances. Nitrogen is thus inert and does not react with other substances very readily. It does not burn with the oxygen of the air. It is, in consequence, used as an 'inert atmosphere'; instant coffee, for example, is sealed in the jar with nitrogen so that the coffee does not oxidize in storage.

Oxygen, on the other hand, has unpaired electrons anxious to form bonds. For living animals, the vital ramification of this is that oxygen uses these electrons to bind to iron atoms in our haemoglobin and be carried by the blood to the body cells.

Even simpler are the rare or inert gases: helium, neon, argon, krypton, and xenon. Each has the magic number of electrons to provide stability round its atomic nucleus. More so than nitrogen, these rare gases are very unreactive and, with the exception of helium, were first discovered as minor constituents of air. (Helium was first discovered through analysis of the sun's spectrum: lines due to the element were observed in the light from the sun.)

Gases have no regular structure. The molecules are in constant chaotic motion, with lighter molecules tending to move faster than heavy ones. At room temperature, for instance, nitrogen molecules move at about 1,000 miles per hour.

If we heat a gas, it will expand—just as solids do; if we prevent expansion by heating the gas in a closed container, the heat energy is taken up by the molecules, increasing their average speeds and thereby causing the temperature to rise. Similarly, if we cool a gas, the molecules move more slowly and the temperature drops.

Intermolecular forces

A substance will be gaseous if the forces which hold its constituent molecules together are not strong enough to overcome the buffeting they receive in collisions with each other. Between gaseous molecules the forces are weak. They are stronger in liquids and even stronger between the molecules or atoms of a solid. The strongest forces are the covalent bonds that result from sharing electrons—as within a single water molecule, binding the oxygen to the hydrogen atoms. In some solids we also meet electrostatic attraction between ions of opposite charge—as in sodium chloride (Na^+Cl^-). (Ions are atoms or molecules which have lost or gained a negatively charged electron and are hence themselves positive or negative.) Sodium (as Na^+) that has lost an electron and chlorine (as Cl^-) that has picked one up together achieve the stable rare-gas electronic arrangement, with eight electrons in a filled shell around the nucleus. Positive and negative charges attract each other very strongly—while like charges repel each other—similar to the north and south poles of magnets.

Sometimes, in a molecule, there may not be a charged ion but an uneven distribution of electronic charge, resulting in a partial positive charge at one end of the molecule and a partial negative charge at the other. This is termed a dipole. These permanent dipoles, resulting from the uneven spread of electrons round the positive nuclei, lead to intermolecular attraction, but this attraction is obviously not so strong as ionic attraction.

A rare-gas atom has no dipolar arrangement, but because of the motion of the electrons round the nucleus, there will be a transitory, instant dipole; at any instant, there are likely to be more negative electrons on one side of the nucleus than the other. Although over time this dipole will average to zero, the effect is none the less to give a small attraction between rare-gas atoms. The more the number of electrons, the greater these weak attractive forces will be, so that, in the He–Xe group of rare-gas atoms, the attraction increases with weight. None the less, these forces are so weak, even in xenon, that all the rare-gas elements are gaseous under normal conditions.

Liquefaction

If we cool a gas, there comes a point where the attractive forces between the molecules exceed the repulsive forces that result from random bumping—as this bumping is less and less strong the slower the molecules move—and the gas liquefies. This process can be aided by increasing the pressure, thereby forcing the gas molecules closer together and allowing the intermolecular attraction to have more effect. Heat is given out when a gas liquefies; conversely, heat is absorbed when a liquid evaporates to a gas.

This principle is used in refrigeration. Refrigerators use a liquid such as ammonia (NH_3) with a low boiling-point (below 0 °C). Evaporation of the liquid requires energy so heat is absorbed and cooling is achieved. The gas so formed is then compressed back into a liquid (this process requires energy), and the heat produced is usually given out at the back of the refrigerator. The ammonia is constantly recycled.

The temperatures we get in the refrigerator depend on the boiling-point of the refrigerant. For the lowest possible temperatures, we require the substance with the lowest possible boiling-point, that is to say with the smallest attractive forces between its molecules. That substance is helium. Liquid helium boils at −269 °C, so that cooling to liquid helium temperatures brings substances close to the absolute zero of temperature, −273 °C.

As substances are cooled, their structures become more and more ordered, because molecular motion is reduced. Gases become liquids which become solids until, close to absolute zero, near total orderliness prevails. Thus, at liquid helium temperatures, we may anticipate some special effects; these are mentioned in the next chapter.

Solutions of gases in liquids

Gases dissolve in liquids. This process is vital to fish, who rely on oxygen dissolved in the water in order to breathe. The dissolving of gases in water is aided by increasing the pressure of the gas and by cooling. We can remove dissolved gases from water by boiling the water, and we can help this process by reducing the air pressure above the boiling liquid with a vacuum pump. Glasses of fizzy lemonade or soda or champagne go 'flat' if left, because when these liquids have been poured from their pressurized containers, the dissolved gases are released.

This dissolving of gases in water solutions creates problems for the human diver. A diver experiences high pressures owing to the weight of water above him; therefore, as his depth increases, the gases in the air he breathes become increasingly soluble in his body fluids. If he surfaces too quickly, then the pressure is rapidly reduced and gaseous nitrogen may come out of solution in the form of bubbles. These bubbles give rise to the condition known as 'the bends', which is dreaded by divers.

One technique for minimizing this problem is to keep divers in sealed decompression chambers when they surface and to reduce the pressure slowly—at a rate which avoids spontaneous bubble formation. An alternative molecular technique is to let the diver breathe not a mixture of oxygen and nitrogen, as in normal air, but a mixture of oxygen and the much less soluble helium. One comic but quite serious drawback of this technique is, however, that the diver, when talking, sounds rather like Donald Duck, and communication becomes difficult. The origin of this effect is the mechanism by which speech is produced: the air in the vocal cords vibrates, as in musical instruments; if the gas is changed from our normal air mixture, the velocity of sound, and hence the frequency or pitch, is altered, resulting in a honking sound. Choirboy pranksters have been known to exploit this effect by filling organ-pipes with helium.

General anaesthesia

A further problem encountered in diving is that when the pressure becomes very high, quite simple gases, including nitrogen, cause first narcosis and, ultimately, anaesthesia. Many of the effects of quite small molecules on living systems are (as we shall see when we discuss drugs) extremely subtle and caused by minute amounts. General anaesthesia seems to be a much cruder affair. Any gas, given sufficient pressure, seems capable of producing anaesthesia; but the greater the gas's solubility in fat, the greater its effect.

The nerve cells of the body, as indeed all cells, are surrounded by a membrane which is mostly lipid or fat. It seems that by disrupting the membrane of the nerve cell, the anaesthetic gas molecules prevent messages of pain being passed to the brain, allowing us to sleep peacefully while the surgeons get to work.

For clinical use, it is necessary to have gases which are anaesthetizing at normal pressures. Two of the commonly used varieties are chloroform and halothane (molecules containing carbon, hydrogen, and halogen atoms). All have relatively simple structures and most are not very reactive chemically. Their use as anaesthetics has had a dramatic effect on medicine, making our own century, surgically speaking, a preferable time in which to live.

Further research

Gases are studied by physical chemists in an attempt to determine precisely the nature and magnitude of intermolecular forces. Were it possible to predict accurately the forces between molecules, then many of the properties of substances would be calculable using computers, eliminating the need for experiment.

From a knowledge of the forces between two water molecules, it is possible to do a computer simulation of liquid water. Although massive amounts of computer time are required, the results of these computer simulations are impressive. Most of the observed properties of water can be predicted. This type of study is now being extended to other substances, and computer simulation of chemical problems is undoubtedly going to be a major research activity in the next decades, possibly using purpose-built computers.

4

The earth's crust—solids

The chemistry which led from the simple molecules of the early universe to the solid structures which make up stars, planets, and our own earth had millenniums in which to create relatively complicated molecular products. Most of the substances available to primitive man were so complex that, as mentioned, many centuries of almost futile study by alchemists were necessary before any real understanding could emerge, which is why chemistry has been a late developing science. In particular, the substances which make up the earth's crust—rocks, soil, and minerals—are generally three-dimensional arrays of atoms or molecules, more variable than the one-dimensional chains of molecules found in living systems. Just occasionally, however, nature does provide almost freak arrangements of atoms which are particularly simple.

Diamond

Diamond, which is composed solely of carbon atoms, is structurally one of the simplest solids and is of considerable intrinsic interest. As we saw in Chapter 1, carbon needs to form four chemical bonds in order to achieve a stable grouping. The simplest example of such a grouping is methane (CH_4), which is a gas found sometimes in coal-mines and as a constituent of natural gas. A special property of carbon atoms is that they also tend to form shared-electron bonds with other carbon atoms. Each of the hydrogen atoms in methane can thus, in principle, be replaced by a carbon atom, which can in turn be joined by shared-electron bonds to three more, and so on (as in Figure 5). The resultant giant molecule—diamond (shown in Figure 6)—is so heavy that it is solid rather than gaseous, and although it occurs naturally, it

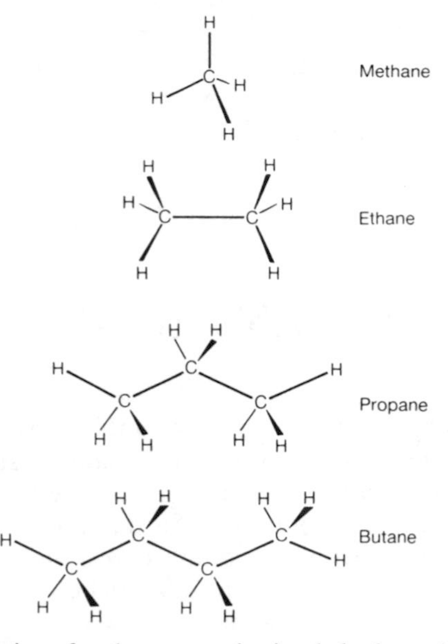

Figure 5. The chains of carbon atoms in simple hydrocarbon molecules

does so rarely, probably because of the tremendous pressures needed to form the solid.

The hardness and strength of diamond derive from the fact that there are strong chemical bonds in every direction. On the one hand, it is very difficult to crush diamond. On the other hand, it can be cleaved along planes which have fewer chemical bonds, a property utilized by jewellers, and there are channels in the structure through which it is possible to pass small particles such as electrons. Models such as those in Figure 6 provide an excellent illustration of how the macroscopic properties of a substance are related to its structure at an atomic or molecular level.

Mention should also be made of graphite, another form of pure carbon. Here the carbon exists in loosely bonded layers, and hence the material is much softer than diamond.

Common salt

Another well-known solid substance with a very simple structure is common salt (NaCl), with one sodium atom (Na) for every chlorine (Cl). In this case, the atoms are not bonded to each

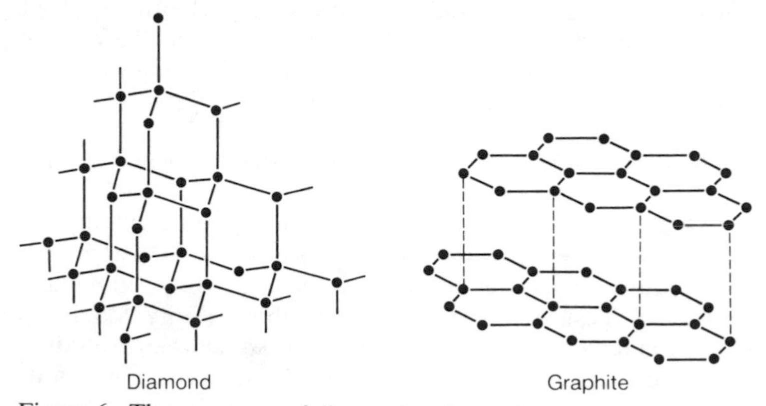

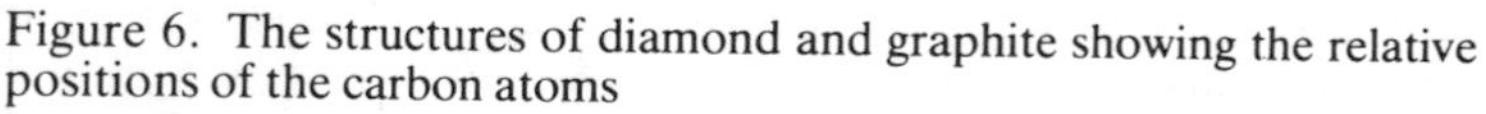

Figure 6. The structures of diamond and graphite showing the relative positions of the carbon atoms

other by sharing electrons but exist as ions (Na^+ and Cl^-). Ions, as mentioned previously, are atoms or molecules which have lost or gained a negatively charged electron and are hence themselves positive or negative.

A stable crystal of salt (sodium chloride) consists of a lattice of alternating ions. Again, the properties of the macroscopic solid can be predicted from this model. The electrical attraction between the ions is very strong, since they are so close together, and therefore substances like salt have very high melting-points: we have to put in a lot of heat energy to disrupt the structure and make the atoms flow. But if we surround the ions in the salt crystals with water molecules, we shield the strong electrical interactions between the positive and negative ions and the crystals break up, due to the reduction in interionic attraction. Consequently, although it is hard to melt salt, it can be dissolved very readily.

Metals

A metal may be envisaged as an orderly array of positively charged ions embedded in a sea of electrons. The overall structure is electrically neutral, with the sum of the positive charges on the metal atomic ions being equal to the number of relatively free and mobile electrons. The electron sea accounts for two of the most obvious properties of metals: the conduction of electricity and of heat. The electric current in a metal is caused by the free, negatively charged electrons flowing towards the positive pole of a battery or mains supply. The heating effect of the electric current (as in an electric fire) is caused by the flowing electrons bumping into atomic nuclei which are not mobile; the resulting kinetic energy of the nuclei being manifest as heat.

Metals comprise a large proportion of the chemical elements. Gold has been known from prehistoric times. It is malleable and can be hammered into shape, causing one layer of atoms to slide over another. Harder metals can only be worked at higher temperatures when the interatomic forces are reduced by heat input, which causes the atoms to shake about. Purity, too, influences a metal's hardness. Pure iron is workable when heated, as our ancestors found in the Iron Age. By contrast, pig-iron or cast-iron is extremely hard and brittle. Such iron comes straight from a furnace and contains a small amount of carbon from the coke used. The carbon atoms are randomly distributed in the lattice of iron atoms and prevent planes of atoms sliding over one another.

Most metallic elements do not occur naturally in their pure form. Only gold, silver, mercury, and occasionally copper do so. Many occur as constituents of minerals in which they are combined with a variety of other elements; but when the minerals are heated, some metals are easily oxidized by the atmospheric oxygen to yield metallic oxides. Extraction of the metal then proceeds by removing the oxygen from the oxide—by heating the oxide in a furnace together with carbon in the form of coke or coal. The metals produced in this way include the commercially important iron, copper, lead, tin, and zinc. The extraction of these metals has had profound effects on human history, resulting in the Iron Age and, later, the Bronze Age (when it was discovered that mixing copper and tin produces the alloy bronze).

Other metals which form more strongly bonded oxides than, say, copper and iron were not prepared as pure elements until the last two hundred years or so. These included sodium, magnesium, and aluminium. Compounds containing these elements are very abundant in the earth's crust, but the extraction of sodium metal from sodium chloride and of the other elements which readily form positive ions from their compounds had to wait until electrolytic methods were available (that is, until the electric battery was invented at the beginning of the nineteenth century). When an electrical current flows through melted sodium chloride, the positive sodium ions move to the negative plate, or cathode, of the cell (where they form the metal) while the negative chlorine ions move to the positive anode (where they are given off as chlorine gas). The light metal aluminium is very abundant in the earth's crust as an oxide called bauxite. The extraction of this very useful metal, so important in aircraft production, demands massive amounts of electricity to electrolyse the aluminium oxide. For this reason, aluminium production is largely concentrated in regions where there is abundant, relatively cheap electricity.

The cost of the electricity or other energy used is a major feature in the production of metals and, indeed, throughout chemistry. Always the chemist is constrained by the laws of thermodynamics, the first of which essentially states that you cannot get something for nothing.

Glass

Although glass appears solid, at the molecular level it resembles a liquid. In fact it is an amorphous solid, a description which provoked from Erwin Schrödinger, the father of quantum mechanics, the aphorism that 'so-called amorphous solids are either not really amorphous or not really solid'. Thus the constituent molecules are not laid out regularly in a lattice but are more disordered. The mineral quartz, from which glass is manufactured, is made up from silicon and oxygen atoms, with two oxygens for each silicon (SiO_2). In the quartz crystal, the SiO_2 molecules form helical chains. If quartz, an impure variety of which is sand, is heated above its melting-point and then cooled rapidly, the silicate chains do not have time to return to a perfectly crystalline array. They harden into a disordered structure—glass. Special forms of glass are prepared with oxides of lead, boron, aluminium, sodium, or calcium mixed with sand. Elements like sodium or calcium form ionic bonds with the silicate oxygen atoms, giving a softer glass with a lower melting-point. Boron and aluminium form shared-electron bonds to produce pyrex glass, which is almost as strong and heat-resistant as pure silicate glass or fused quartz, which also has a complete network of covalent bonds.

Novel solids

The contemporary world is being transformed by the silicon chip and other 'solid-state' electronic devices. The materials used in these devices are relatively obvious: metallic elements act as conductors, silicon and simple magnetic materials as memory elements. Nature, by contrast, uses organic molecules to create the even more impressive machinery of the human brain. It thus seems attractive to chemists to devise better materials, possibly learning from nature, to replace substances now incorporated into electronic devices. The search is on for organic materials which will act as conductors of electricity, semiconductors, memory elements, and storage devices for data and optical displays.

Success in these challenging areas seems most likely to come from combining the skills of the organic chemist, who can design molecules with specific properties, with those of the polymer chemist, who can make plastics and similar materials.

5

The earth's rivers and seas—liquids

The temperature at which a gas becomes a liquid depends on the strength of the attractive forces between the molecules. Generally, as we have seen, we expect the intermolecular forces to increase with the complexity of the molecules and the number of electrons these possess. At room temperature, simple molecules such as carbon dioxide, ammonia, and hydrogen sulphide are gaseous while more complex structures like benzene or acetone (nail varnish remover) are liquids. In other words, the kinetic energies possessed by the simple molecules at room temperature mean that the intermolecular forces cannot prevent the molecules separating from each other: a liquid would just evaporate. Bigger molecules are held together more strongly and will form a liquid with a surface. Heat is then required to evaporate the molecules, taking them from the liquid to the gaseous phase.

Water

The one glaring exception to the above generalization is water: water, composed of light and simple molecules (H_2O), is liquid at normal temperatures and has to be heated to boil. This must mean that for some special reason there are strong forces attracting water molecules to each other; and there are indeed special sorts of bonds which form between molecules containing hydrogen atoms bound to one of the three elements—nitrogen, oxygen, and fluorine. The nuclei of these particular elements hold their electrons so tightly that in the formation of shared-electron bonds with hydrogen atoms, they end up with a disproportionate share of the electrons; consequently their atoms carry a partial negative charge and the hydrogen atoms a partial positive charge.

The attractive forces between the partially charged molecules result in what are called 'hydrogen bonds'. They are particularly important between water molecules.

Hydrogen bonds not only hold water molecules together to form a liquid at room temperature but also have a tremendous importance in biology and life processes. The DNA chains of genetic material are held together by hydrogen bonds. The bonds are also of crucial importance in producing the required three-dimensional structures of enzymes by forming cross links between the chains of amino acids, as we shall see later. They add rigidity to the structure. Figure 7 gives an idea of the structure of water, with hydrogen atoms represented by open circles and oxygen by solid circles.

In liquid water, the individual water molecules (H_2O) each form hydrogen bonds between the H of one molecule and the O of another. The liquid does not then consist of chaotically moving single molecules; there is some short-range order, with the position of one molecule depending on the location of a neighbour to which it is joined. A consequence of this ordered structure of molecules is that the solid form of water, ice, has a more open structure than the liquid; therefore close to freezing-point (below 4 °C), water is more dense than ice. Hence ice floats on water. This fact is so familiar that it is hard to realize its significance. In almost no other instance does the solid form of a substance weigh less than the liquid. If our planet were covered, as are others, by ammonia, then things would have been very different, and the world would not have evolved in the way it has.

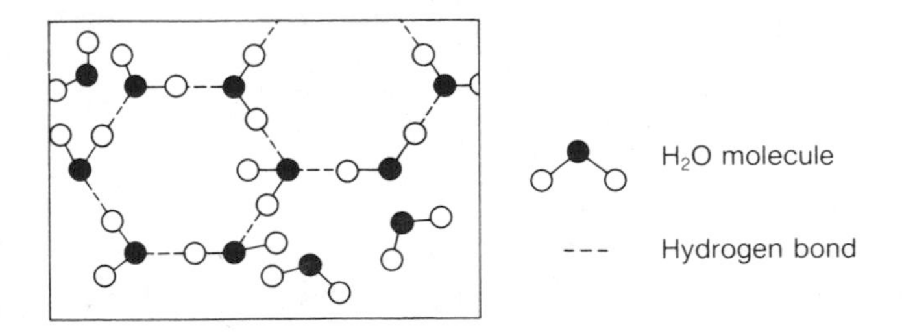

Figure 7. An instantaneous view of the structure of water

Each H_2O molecule may be joined to others by hydrogen bonds, which form, break, and re-form.

Again, because of the strength of the hydrogen bonds holding water molecules in the liquid form, a lot of heat energy is required to break the bonds and vaporize water. This property helps to control our body temperature—by evaporating water as sweat. Because of the strength of hydrogen bonds, this evaporation requires a lot of heat energy for only a small quantity of water. The heat comes from the human body, thereby producing some cooling.

Aqueous solutions

As well as being the most abundant substance on the earth's surface, water is also the most widely used solvent. Solids are dissolved in water to give aqueous solutions. Water again owes this property to the fact that the oxygen atom attracts the electrons in the O–H bond very much more strongly than does the hydrogen atom. In consequence the oxygen atom is somewhat negatively charged, having more electrons close to it than the number of protons in its nucleus, while the hydrogen atoms each carry a partial positive charge. This results in the molecule having a dipole moment, negative at the oxygen atom and positive at the hydrogen ones. Not only does this dipole contribute to the hydrogen bonds between water molecules, it also helps water molecules to bind to ions, such as Na^+ and Cl^- in common salt, which bear whole charges, or to other molecules with dipolar charge distributions. Water is thus a particularly good solvent for these so-called polar substances. By contrast, substances formed from molecules which are uncharged and do not have dipolar charge distribution are insoluble in water: the most obvious example is oil, made up from hydrocarbon molecuies containing only hydrogen and carbon atoms, neither of which have a strong attraction for electrons.

Lubricants

Lubricating oils consist of atoms joined in long, chain-like molecules with, consequently, large inter-molecular forces. It is hard to push something through such liquids because the molecules tend to stick together and to become entangled. At the macroscopic level, the liquids are viscous and will stick to a metal surface, thereby providing a lubricating layer.

A particularly curious liquid is ethylene oxide. In water, ethylene oxide reduces the viscous drag of water attraction and enables one to speed up a boat if, for example, it is sprayed on the hull. As such, it has had to be banned from racing yachts. It is, however, used in fire-hoses to reduce the friction between water and the walls of the hoses.

Soaps and detergents

Although soaps and detergents are usually obtained as solids, it is in solution that their properties are so important. They are molecular tricks to enable substances which are insoluble in water, such as oil and grease, to be dissolved. The basis of the mechanism (see Figure 8) is to have a detergent molecule with one end a hydrocarbon, hence soluble in the greasy substance, and the other end the charged group of an acid, soluble in water. As the detergent molecules surround the oily dirt, dissolution is achieved. The hydrocarbon attaches to the dirt, surrounding the water-insoluble molecules, while the other end of the detergent molecule permits solubility.

The influence of detergents in reducing the drudgery of washing is just one of the many ways in which the chemical industry has contributed to a better life in the home.

Liquid helium

If water is the most important liquid, liquid helium is the most surprising. The intermolecular forces (or, in this case, interatomic forces, since helium atoms are stable as single entities) are very

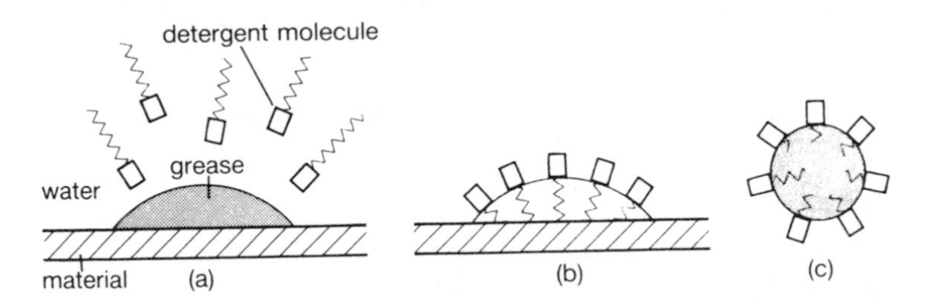

Figure 8. The action of detergent molecules on grease

(a) Some material with grease on its surface is immersed in a solution of detergent. (b) The tails of the detergent molecules dissolve in the grease but the heads are attracted to the water. (c) The grease is surrounded by detergent molecules and is therefore now soluble in water.

small because the atom has only two electrons. Hence, of all the elements, it requires the lowest temperature to cause liquefaction, and is used as a refrigerant when the lowest possible temperatures are required—as, for example, when studying superconductivity. This is a curious and potentially important phenomenon where electrons move in pairs, not repelling each other as they usually would, and metals have no resistance. With no resistance, huge currents can flow. The first practical application of superconductivity has been to produce electromagnets with massive magnetic fields, now used in new forms of body scanners without X-rays.

At extremely low temperatures, approaching absolute zero, when one might expect helium to solidify, it in fact remains liquid. However, the structure is extremely ordered. Ordering at low temperatures is predictable from the second law of thermodynamics which states that at high temperatures systems become more disordered and chaotic. This ordered array of helium atoms gives rise to some surprising phenomena. If, for example, the liquid starts to flow, it will continue flowing, all the atoms moving regularly in phase with each other. In this way, the liquid can flow out of a beaker by going up and over the walls of the vessel! The demands of order outweigh the constraints of gravity.

Liquid crystals

There is a class of liquids which share some of the features of solids and are therefore named 'liquid crystals'. Essentially, they consist of rod-like molecules which pack together in such a way as to have an ordered structure in one dimension and a random structure in another. Particularly important from a commercial point of view are structures where the molecules form layers, the apparent colour of the liquid crystals being highly dependent on how far apart the layers are and on the angles between the molecules in adjacent layers. Changing these variables either by altering temperature or by mechanical pressure causes a colour change in the crystals.

The applications of these sensitive colour changes are manifold. One can buy liquid crystal thermometers based on digital patterns of liquid crystals which are colourless except at the appropriate temperatures. Liquid crystal materials applied to the body can be used to investigate organs which radiate heat at a different rate from surrounding tissue, or tumours, or the severity of burns. The colour of a liquid crystal tape can be used as a constant monitor of an infant's temperature.

Perhaps the greatest potential of liquid crystals is in display devices—for example, digital watches. One way of achieving a display is to polarize the long molecules of the liquid crystal solution so that they are positively charged at one end and negatively charged at the other, and to dissolve in this solution an ionic substance which possesses an electronic charge. When placed between the plates of a condenser (one plate positive and the other negative), the solution becomes cloudy because the aligned long molecules are interfered with by the small ions pushing past them towards the oppositely charged plate. On switching off the condenser, the orderly structure immediately returns, the turbulence disappears, and the solution becomes clear once more. The ability to switch an area of liquid crystal between clear and cloudy is the basis of display devices.

The chemist's contribution in this area is, so often, to devise molecules which have the required properties for a particular application.

6

Molecules and life

One of the most fundamental and perplexing questions in science is how life began. Although some people believe that the origins were extraterrestrial or mystical, the more conventional scientific view is that life developed here on earth: on an earth which is itself derived from the molecules first formed in the period following the big bang. The solid matter of the earth consists of three-dimensional structures of atoms and molecules; living systems more often consist of one-dimensional chains of molecules and, in this respect if in no other, are structurally simpler.

Replicating molecules

Chemists have reproduced in their laboratories the conditions which seem likely to have prevailed on the primitive earth—sometimes described as the 'primordial soup'. It is clear that electrical storms or ultraviolet light from the sun could have provided the energy to break down molecular structures, allowing them to re-form in many possible arrangements, and experiments have indeed demonstrated that those organic molecules which are the important building blocks of living systems are produced by these supposed conditions.

Producing the constituents of living materials is, of course, a long way from creating life, however significant a first step it may be. Life in a primitive form will have been created only when the molecules can reproduce or are self-replicating. Given the millions of years that such a process had in which to evolve, it seems not unlikely that the events may have occurred in the following way.

The carbon atoms in organic molecules can frequently be induced to join on to each other to form a chain. If each of the molecular building blocks in a chain were specifically attracted to a similar molecular building block, a copy of the original chain would develop, with the two strands stuck to each other. Each chain thus becomes a template for a new copy of itself. If the pair of weakly bonded chains then separate, each single chain can be responsible for the reproduction of another identical copy, continuing until the supply of molecular building blocks runs out.

This replication of molecular chains represents in essence the behaviour of our own genes. Genes are chains of four basic building blocks (nucleic acids), the structural nature of which (the double helix) was described by Watson and Crick in 1953—one of a series of successful interventions into biology by chemists and physicists.

Evolution

Inevitably, as in any copying process, there will be occasional mistakes, and therefore minor variants of the parent structure will be produced. If these variants are more stable under the prevailing conditions, then chance alteration will have evolved a better-adapted species of molecular chain. This simple process is sufficient, given the time which has elapsed since the earth was formed, to provide an understanding of evolution.

As a result of this evolutionary process (or so we may speculate), the replicating molecule has become the deoxyribonucleic acid (popularly called DNA) which constitutes the genes of contemporary organisms: organisms which can be viewed as survival machines evolved for the benefit of these gene molecules. Much of evolution may be understood in terms of the replicating molecule or gene adapting to ensure its survival. The phrase 'survival of the fittest' begs the question 'fittest what?'. Rather than fittest species or individual animal, the fittest gene is the chemist's answer. Geneticists also take this view, and the phrase 'the selfish gene' has become a cliché.

DNA is made up from a series of individual small molecules usually called bases. The DNA molecule acts as a code which controls the production of proteins by living cells. The list of DNA bases is read as a code for the building blocks in the protein. The proteins are themselves one-dimensional chains; but in the protein chain, each link comprises one of about twenty molecules of a type called amino-acids rather than one of four

DNA bases. If there were only four types of amino-acid, then the DNA code would be a simple one-to-one translation; each of the four DNA bases would specify or 'code for' one amino-acid. If the code required a pair of DNA bases to stipulate one amino-acid, there would be the possibility of 4×4 different interpretations. A triplet code with $4 \times 4 \times 4$ possible sets of three consecutive DNA bases thus has more than enough possibilities to code for twenty different amino-acids and also for some additional features. Three consecutive bases stipulate each amino-acid in the protein. This pattern is essentially the same for all living systems, be they viruses or human beings. There is a temptation to describe the viruses as simple and to think of ourselves as superior, but in evolutionary terms the replicating molecules of both 'extremes' have survived to the present. The 'higher' organisms have developed sex as a way of mixing genes and thus can avoid some of the pitfalls of reproducing damaging characteristics. But for all forms of life the replicating molecule is what is passed down the generations, and this replicating molecular chain controls the production of the all-important molecules, the proteins.

Proteins

Proteins are the molecules which perform most of the jobs required by living cells and organisms. The DNA of the genes provides a blueprint which specifies the sequence of the amino-acids which make up the protein, as in Figure 9. The actual sequence predetermines just how each long protein chain will fold up. Some of the amino-acids have a tendency to dissolve in water; others shun it, causing the chain to fold into a globular shape whereby some parts of the chain are shielded from the water which surrounds the protein in a living cell. These three-dimensional protein shapes are far from random, and are designed to accomplish specific tasks at a molecular level.

Protein molecules provide the machinery and much of the structure of animals and plants. Some proteins have the function of breaking down other proteins—as in the chemical reactions for digesting food. Some, called enzymes, are used in the conversion of molecules from one form to another, creating the molecular substances used in the body's chemistry. Yet others act as regulators in the production of chemicals, or carry the

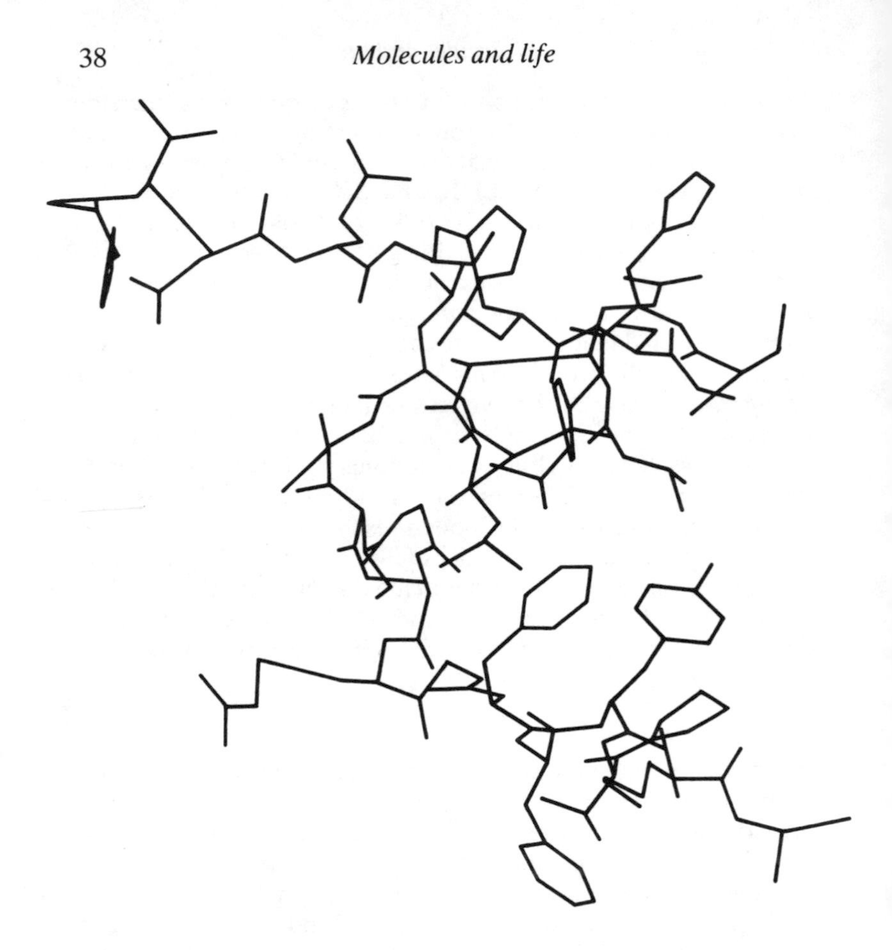

Figure 9. The folded structure of the protein chain insulin

oxygen in the blood from the heart to all the organs in the body.

Because these very fundamental processes are common to all living systems, the proteins which perform the processes tend to be similar but may exhibit minor variations from species to species. The more similar the species, the closer in structure are their proteins. In this way, we can see at the molecular level how evolution has been accompanied by slow changes in some of the amino-acids in the protein chains.

If we look at a certain protein in the horse and the donkey, for example, we see that there is only one minor change in a single amino-acid amongst several hundred which remain the same in the two species. Thus horse and donkey have to be very close in evolutionary terms—indeed, they can mate and produce off-spring—and it is possible to inject the protein of one into the other without damage. Between more distant species, injections of foreign proteins cause damage—for example, organ rejection. Hence kidney transplants between twins or closely related individuals are more likely to be successful than those between unrelated people. The actual differences between people can in some instances be understood in molecular terms, as in blood groups.

Molecular diseases

A mutation or change in the DNA sequence of a genetic message will cause a change in an amino-acid in the protein for which the DNA provides the prescriptive code. Throughout time, this has been the basis of evolution. Although some changes are damaging, these tend to be less serious for human beings (or any other organisms that reproduce sexually) because with two copies of each gene (one from each parent) and with only one alternative being expressed in the formation of the protein molecule, there is a good chance that the damaging change will not be represented in the offspring. But if both DNA copies prescribe the same defective protein, then the unlucky individual will have inherited a genetic, or molecular, disease. Haemophilia is an obvious example, a disease which became a serious problem for European royal families, who tended to intermarry.

This type of problem has been particularly studied in the case of haemoglobin, the protein which carries oxygen in the bloodstream. A single change in an amino-acid may produce haemoglobin molecules which stick together. This makes an observable difference in the blood cells and results in the disease called sickle-cell anaemia. A similar genetic alteration gives rise to other defective haemoglobin molecules and to distressing diseases such as thalassaemia.

Since the DNA molecules coding aberrant protein molecules are passed from generation to generation, the occurrences of these molecular diseases are to an extent localized. In particular, sickle-cell anaemia is found amongst Negroes and thalassaemia in people of Middle Eastern origin. The characteristics have probably survived because, despite being deleterious (or not 'the fittest'), they may provide resistance against such diseases as malaria.

Manipulating evolution

Now that these genetic diseases are understood at the molecular level, it is tempting to use our knowledge to alleviate or prevent them. The most obvious way to do this is to use a drug which, in the case of sickle-cell anaemia, prevents the haemoglobin molecules sticking together. The drug molecules have to be small and be able to bind on to the large protein molecules, preventing their association.

Prevention is also a possibility since some of the crucial protein molecules from the womb of a pregnant woman can be extracted with a fine needle and tested as to whether the foetus has inherited the condition (know to be a possibility from its family history). Based on this knowledge, the decision on whether or not to abort the foetus may be taken.

This is the first of many examples in this book where knowledge of what is happening at the molecular level leads very dramatically and simply to profound social, moral, and economic consequences. Here the knowledge that one amino-acid out of the 574 in the haemoglobin molecule is a variant from the normal can result in the abortion of a child.

An even more dramatic interference with evolution has come about with the advent of what is popularly called 'genetic engineering'. It has been found possible to cut out lengths of the DNA genetic message of one species and insert it into the code of another. As yet, this is restricted to persuading bacteria to manufacture proteins outside their normal range, and has opened up the exciting possibility of using bacteria to produce, for example, the insulin needed by diabetics. Diabetics do not produce enough of the protein insulin, and in consequence they have to inject themselves with insulin derived from sheep, similar but not identical to their own. If the chosen bacterium could have appropriate genes added to cause it to manufacture insulin of the human variety, then diabetics would face fewer problems.

As well as opening up potentially important commercial

possibilities, this advance also poses moral and social questions. If man is not actually creating life, new forms of life are being created, and evolution is being manipulated. But perhaps that is our nature.

Research activity

The chemist (or more precisely the biochemist and molecular biologist) has developed ways of determining the sequence of the nucleic-acid building blocks of DNA and, either independently or as a consequence, can work out the linear sequence of amino-acids in protein chains. This work is particularly associated with the Cambridge chemist Fred Sanger, winner of two Nobel prizes. But knowing the linear sequence of amino-acids in a protein molecule is insufficient to determine the overall three-dimensional shape of the coiled-up chain. This shape is an essential part of the mechanism of the protein if it is, for example, an enzyme. Trying to predict three-dimensional structure from a one-dimensional sequence remains an unsolved problem.

Since the three-dimensional structures of molecules like proteins cannot be predicted, the experimental approach is employed, with ever more subtle techniques being devised to extract information. X-ray crystallography has long been the most fruitful approach, but newer methods such as nuclear magnetic resonance spectroscopy (see pp. 91–2) are now making major contributions.

The chains formed from nucleic-acid bases and from amino-acids have been much studied and are well understood. Attention is now turning increasingly to a third group of biological macro-molecules (molecules containing a very large number of atoms): the polysaccharides: consisting of chains of individual sugar molecules, which are small rings of carbon and oxygen atoms. The polysaccharides have important structural properties in biology; and to the chemist, they have the added fascination that as well as having simple linear chains, branching may occur, giving a whole new variety of possible three-dimensional structures, rather like networks or cages.

So successful have the contributions of chemists been to biology that there has been a perceptible shift in the direction of much chemical research—towards the problems of biology.

7

Symmetry in chemistry

Generally, chemistry is a down-to-earth, even materialistic subject. Research problems frequently arise from a response to some need. It may be that a substance with a new quality is needed—a dye which will not fade or a container which will not melt. Nevertheless, chemists do become involved in the deep and fundamental problems which are at the heart of science and pervade all disciplines. One clear example is the question of symmetry. At all levels of science, from biology through to fundamental physics, this topic is encountered. Human beings are almost symmetric but our hearts are on the left of our bodies and the majority of us are right-handed. Some plants always coil to the left while others always coil to the right. These biological properties are not random.

Molecules may occur in equivalent forms except for the fact that they are mirror images of each other—in other words, there are right-handed and left-handed molecules. 'Handedness' is one aspect of a very important branch of chemistry known as stereochemistry—important because the positions of the atoms in a molecule (the molecule's spatial configuration) often have a profound influence on how that molecule reacts (its reaction mechanism) and how fast it reacts (its reactive rate).

Left- and right-handed molecules

Many molecules, particularly organic molecules (or those containing carbon atoms), may be described as 'left-handed' or 'right-handed'. Even in very simple molecules, there is a possibility of handedness occurring. The carbon atom generally joins to four other atoms or groups of atoms in a three-dimensional tetrahedral shape, as illustrated in Figure 10. If the four groups

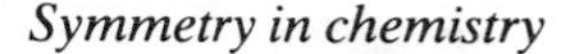

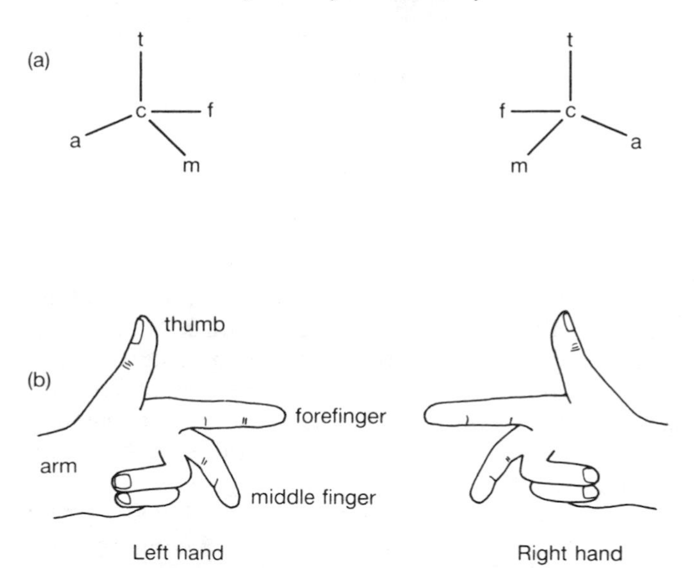

Figure 10. Handedness of the carbon atom when joined by four different groups or atoms

(a) The tetrahedral carbon joined to four different groups labelled a (arm), t (thumb), f (forefinger), and m (middle finger) in the two optical forms. (b) The parallel forms of left and right hands.

attached to a carbon atom are all different, they can be disposed in either a left- or a right-handed fashion. (This effect parallels the rough left- and right-handed tetrahedrons made up from the arm, thumb, forefinger, and middle finger of two hands.) If any two groups are the same, then a given arrangement is identical to its mirror image and cannot be described as a left- or right-handed version (see Figure 11). In the normal laboratory synthesis of compounds with left- and right-handed versions, a process involving many millions of molecules in order to have weighable quantities, the result is an equal production of left- and right-handed molecules. By contrast, nature almost always produces molecules of one unique type. The tartaric acid produced in fermentation consists of just one of the two mirror-image forms, whereas in laboratory synthesis both forms are made in equal amounts.

The actual properties of the left-handed and the right-handed version of the same molecule are in most respects identical: the

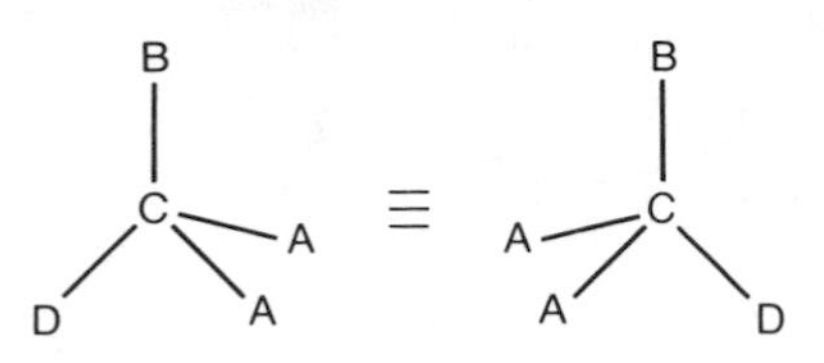

Figure 11. The molecule CA_2BD, which does not exhibit mirror symmetry

same weight, the same chemical reactions. A notable difference, however, is in their behaviour towards polarized light (see Figure 12). Light consists of waves oscillating in three dimensions. The colour of the light depends on the wavelength or frequency of these waves. With a polarizer (such as polaroid sun-glasses), it is possible to create light waves which oscillate all in one plane. This polarized light when shone through a substance which contains molecules of only one handedness has its plane of polarization rotated one way for left-handed molecules and the other for right-handed molecules. What actually occurs is an interaction between the electric field of the light and the electric charges of the positive atomic nuclei and negative electrons in the molecules. The interaction tends to distort the path of the light so that the plane of polarization becomes bent in one direction or the other. If the substance through which we shine

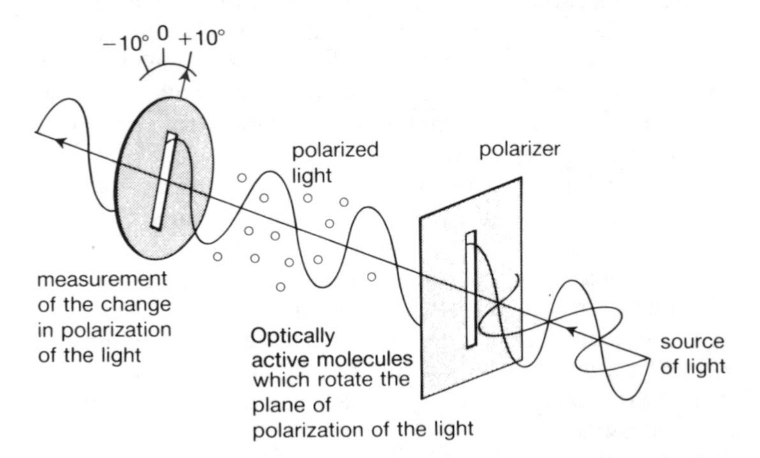

Figure 12. Schematic illustration of the measurement of optical rotation

the light is a laboratory-made mixture of left- and right-handed molecules, then there is no rotation of the plane of polarization: half of the molecules twist the light clockwise while the other half twist it anticlockwise.

Pasteur and handed molecules

One of Louis Pasteur's many brilliant contributions to science was his discovery, based on his study of tartaric acid, of the existence of handedness in molecules. Natural tartaric acid rotates polarized light and is described as optically active; but the mixture of left- and right-handed forms of tartaric acid prepared by chemists is not optically active, despite being identical in most other ways. Pasteur made crystals of both forms of the acid, and found that the crystals reflected the handedness of their constituent molecules. Crystals from grape tartaric acid are all of one form, but the non-active laboratory mixture when crystallized has an equal number of crystals of the two forms, one form rotating light like the natural compound, the other form doing so an equal number of degrees in the opposite direction. Pasteur separated the two kinds of crystals with tweezers and a magnifying glass. In the laboratory, we now have rather more sophisticated techniques for separating asymmetric molecules into their left- and right-handed forms.

Pasteur then discovered that a certain plant mould, when allowed to grow in a tartaric acid solution containing a mixture of the mirror-image forms, only utilized one of the forms; consequently, the solution became optically active. Pasteur perceived great significance in this fact. In so far as the living organism, the mould, was selecting only one of the two mirror-image forms of tartaric acid, he concluded that it was built from symmetric molecules of one particular handedness. Further he went on to postulate that preference was a general phenomenon and a way of distinguishing between non-living matter and matter that either is or has been alive. Subsequent research has confirmed Pasteur's view. The fact, as it now is, can thus be used to answer questions about the existence of life on other planets. Space exploration may bring back samples of the rock from Mars or Venus. The material may yield carbon-containing molecules. If these molecules are of a predominantly left-handed or right-handed nature, scientists would be convinced that life exists or has existed on the planet. So far there is no compelling evidence, for although some of the organic molecules found in meteorites are of only one handedness (and hence of living origin), they may be terrestrial contaminants.

Mirror-image molecules in nature

The most obvious function of handedness relates to fitting shapes, just as with hands and gloves. We can imagine a 'molecular glove' which will accept only a left-handed molecule or only a right-handed molecule. Physiologically we taste molecules if they fit into a molecular pocket which is designed to recognize shape. Molecules which fit a taste receptor molecule will provoke a response which the brain recognizes as taste; those which do not are not tasted. The right-handed form of phenylalanine (an amino-acid) is sweet while its left-handed form is bitter. Conversely, the right-handed form of the amino-acid valine is bitter and the left-handed form tastes sweet. One form of glucose is sweet, while its mirror image tastes salty.

Many drugs work by fitting into molecular pockets designed to accept molecules which pass on nerve impulses or stimulate hormone secretion. As we shall see later the drug molecules often block these receptor pockets. Again the left- and right-handed forms of a given drug may have dramatically different effects because one shape fits the molecular pocket better than its mirror image. An example is amphetamine: one form produces an effect, the other does not. Vitamin C has an effect in only one mirror form: taking doses of its mirror-image molecule will not prevent scurvy in the way that the natural molecule from citrus fruits does. The fact that, in many instances, only one form of an asymmetric tetrahedral shape will bind to a physiological receptor, as the molecular pocket is more properly called, leads to the important conclusion for the pharmaceutical industry that at least three of the corners of the tetrahedron must be involved in binding for the receptor to show any preference. If only two corners bind then both left-handed and right-handed forms could attach themselves.

Even the small molecules involved in living systems not only show handedness but also exist in just one natural form. Above all, the amino-acids, from which proteins are built, are all left-handed. We actually possess an enzyme in our livers which will destroy any right-handed amino-acids that we happen to synthesize or encounter.

Clearly, the nature of life is closely related to the symmetry of the molecules which constitute living systems. What now cries out to be answered is why all life is built from molecules of a definite handedness. This remains an unsolved problem, despite a wealth of speculation. Not only are the small molecular building blocks of one specific handedness; so are the larger structures made from them.

The molecule of heredity—DNA

DNA is the molecule of heredity. It is a long thread-like molecule which has a definite handedness, being in the form of a right-handed helix of two strands (see Figure 13). Helices, like spiral staircases, can wind to the left or the right. The genetic code is provided by the four variable units called bases—A, G, T, and C (short for the small molecules called adenine, guanine, thymine, and cytosine)—which can join in pairs across the strands. As mentioned previously, a group of three consecutive bases in this chain provides the code for a single one of the twenty building blocks (amino-acids) which make up proteins. The backbone of DNA is a string of molecules of the sugar type, linked by bridging phosphate groups. The helix is forced to be a right-handed form because of the particular shape of the repeating sugar block in the long chain. Each of the individual small building blocks or links in the chain may be considered to be a right-handed molecule. The helical DNA is itself coiled in helical fashion, but in a left-handed helix—or superhelix as it is sometimes called.

There is much speculation about and some support for the view that the supercoiling of DNA may be a major part of the mechanism whereby cells differentiate. The question to be answered is how cells express only a part of the genetic message they carry in their DNA. When cells divide, using the DNA complementary strands, each daughter cell has the complete genetic message, identical in all cells. Yet some become liver cells and others brain cells. How are some sections of DNA (genes) switched on or off? The explanation which is becoming increasingly compelling is that not only is the DNA helix itself helically coiled but this superhelix is also coiled, giving a super-superhelix. In the superhelix, only the loosely coiled parts can be

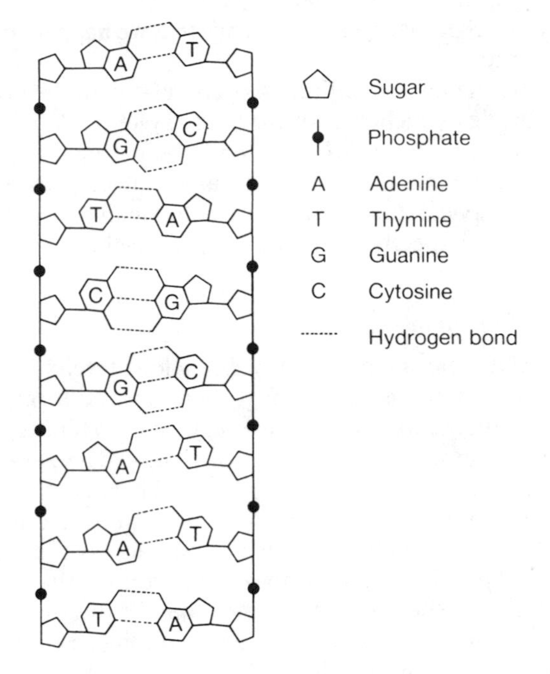

Figure 13. The structure of DNA

replicated, and control is exerted by adjusting the degree of uncoiling. The tightly coiled portions are not approachable by the molecules which read the code by binding on to the DNA.

If the ends of the helical DNA are joined in a circle (as in simple organisms such as bacteria), the superhelical turns are locked in, just like the kink in a Möbius strip (made by taking a paper strip, twisting it, and then joining the ends). The total number of helical turns (n) and superhelical turns (N) is a constant, $n + N$. The sum of these numbers cannot be changed in circular DNA unless the DNA is cut or 'nicked' by an agent such as an enzyme. Right-handed coiling is denoted by positive values of N or n while left-handed coiling is treated as negative turns in the helix.

The DNA superhelix is left-handed (N is negative or less than zero) while the double helix is right-handed (n is positive or greater than zero). Thus if the double helix is uncoiled, n must decrease and N must get closer to zero. Uncoiling the superhelix promotes unzipping of the double helix at certain points, making it ripe for transcription of the genetic code. If enzymes are used to alter the tightness of the DNA supercoils, the rates of transcription are altered. Differentiation of cells through exposure of just some of the genes could therefore be controlled by adjusting the supercoiling.

Excessive supercoiling would so reduce helical coiling that many many genes could be transcribed—a possible mechanism for the runaway growth of cancer cells.

In the higher organisms, where the DNA is not circular but in a series of loops, the supercoils are locked into each loop, with the degree of supercoiling possibly controlled by enzymes. The attractive feature of this configuration is that it suggests how cells may differentiate so long as the degree of supercoiling is inherited by daughter cells: some parts of the DNA message can be read but other sections are buried and unreadable.

An intriguing explanation as to how supercoiling is adjusted involves left-handed helical DNA, which of course has negative turns. This form of DNA can be created in the laboratory but has also been found at specific points in the right-handed DNA of fruit flies. If chemicals can cause a section of DNA to become left-handed, then the supercoiling must be adjusted (if $n + N$ is to remain constant). This adjustment could switch off the genes in a particular loop of DNA.

As yet these mechanisms remain unproven, but it is certain that the interplay of left- and right-handed helices is a big factor in the molecular basis of life.

Origins of molecular handedness

What the Stanford chemist Bonner has aptly described as the original chicken-and-egg problem may be formulated as which came first—life or the use of molecules of only one mirror form? Did life evolve and then gradually come to adopt one form of molecule, or was there a preference amongst the molecules before they developed the facility of replication, reproduction, and life as we understand it? Obviously, once a preference has been created for one of the mirror forms, the nature of reproduction will tend to maintain that choice. Theories which start with the premise that life came before molecular preference are called biotic theories, whereas those which suggest that life developed from an already existing imbalance among the population of molecules on the primitive earth are called abiotic theories.

The simplest but weakest idea is that all life is derived from a single 'Adam' molecule. By chance that molecule will have been either left- or right-handed and all those derived from it will be the same. Other theories look for some mechanism which may have given rise to a preference, such as the polarization of sunlight reflected from the sea. None of the present theories is particularly convincing, but what does seem probable is that handedness in molecules of biological origin depends on asymmetry in the basic laws of physics and their impact on the early universe. The symmetry and asymmetry in nature remain intriguing problems still awaiting an answer.

8

Small molecules in biology

So far, the organic molecules which we have considered have all been large polymers: small molecules joined in chains, with many thousands of constituent atoms. We have met the genetic molecules of DNA; proteins, which provide the molecular machinery in living systems; and also polysaccharide molecules, which act as structural materials. The impression may have been given that small molecules are not involved in biology. Nothing could be further from the truth. While the big polymeric molecules provide the machinery in the form of molecules which do things, the control systems which regulate the actions of the larger molecules utilize small molecules.

Roughly speaking, slow control is achieved through hormones—substances which circulate in the blood or intercellular fluid to those organs subject to their control. Fast control is furnished by the nervous system. Hormones are small molecules; so are the transmitter molecules which pass messages across important junctions in the nervous system.

Hormones

Hormones are produced by living organisms to control the rate of biochemical processes. In plants, hormones control growth, fruiting, and all the seasonal changes such as the shedding of leaves. Much remains to be learned about plant hormones; animal hormones are better understood.

Animal hormones are normally produced by glands. The glands secrete hormone molecules into the bloodstream so that they can be carried to other parts of the body. Some hormones act directly on cells; others stimulate organs to produce their own hormones. Hormone control systems are very sensitive, being equipped to supply feedback and negative feedback. For example, the male sex hormones stimulate the testes, which in turn produce hormones to inhibit the production of the original stimulating hormones. In this way, an optimal balance of hormone is achieved.

The mechanisms of the much studied female sex hormones are even more subtle. If a women takes one of her controlling hormones in the form of a drug, then it may be possible to fool the body's system into reacting as if she were pregnant and so avoid producing ova. The interrelated effects are so complex, however, that it is important to ascertain that there are no side-effects involving related systems. For example, some of the earlier contraceptive pills, whilst mimicking pregnancy, increased the propensity to thrombosis by stimulating the natural protection against excessive bleeding in childbirth.

If, because of disease, there is an over- or underproduction of a hormone, then the consequences can be dramatic and distressing. Goitre and dwarfism are two obvious clinical manifestations involving disruption of the hormone thyroxine. Some of these problems may be overcome medically by administering the appropriate hormone in a purified form—as in insulin treatment for diabetes. Until recently, the insulin always came from sheep, but now chemistry under the name of genetic engineering has been able to alter the DNA of bacteria so that, as mentioned previously, the bacteria can produce molecules which are identical to human insulin molecules.

Closely related to hormones are the pheromones, which are released not into the body of the host creature but into the external world. Pheromones provoke behavioural developments or reproductive responses in recipients that are of the appropriate species, especially in insects. They act as molecular messages between individuals and may control group activity. These molecules, particularly the sex attractants, have been tested as possible insecticides whereby they attract insects to their destruction. In mammals too, there is evidence for inter-animal effects of pheromones. For example, one male in a cage of female mice can cause the oestrus cycles of the females to become synchronized.

The nervous system

Fast control, required, say, when one's hand touches a hot plate, is achieved by a network of nerves. Sensory nerves carry the message to the brain that the hand has touched something hot, and motor nerves carry the message back from the brain to work the muscles for moving the hand.

Each nerve consists of a bundle of nerve cells which, when 'fired' or stimulated, transmits an electrical pulse through sodium and potassium ions (Na^+ and K^+) flowing across the membrane of the nerve cells.

In primitive forms of life, the sensory nerve is connected directly to the motor nerve at a junction called a synapse. If the sensory nerve is stimulated, perhaps by touching something, then the motor nerve is provoked to send the message causing movement. We human beings retain some of this rapid response in the form of reflexes, but in general the incoming information from the senses is co-ordinated within the brain and the spinal cord. The sensory and motor nerves join the spinal cord at points between the vertebrae.

Synapses and nerve transmitter molecules

Nerve cells are elongated; they have a membrane and may be surrounded by a protein sheath which acts rather like the plastic insulation round an electrical cable. The chief difference between nerve conduction and a wiring system is, however, the existence of gaps. Nerves which join muscles and cause them to contract have a gap between the nerve endings and the muscle cells. There are also gaps part way along the nerves and especially between the junctions in the spinal cord and between the nerve cells of the brain.

In molecular terms, these gaps are wide—about a hundred times as big as the size of a small molecule. The transmission of the nerve signal across these gaps is by means of small molecules. When the message reaches the end of a nerve, transmitter molecules are released from the ending, diffuse across the gap, and propagate the impulse by causing a change in the cell on the far side of the gap.

In 1921 Otto Loewi gave the first convincing demonstration that nervous transmission involves chemicals. Working with nerves attached to a frog's heart, he observed that a substance he named *vagusstoff* is liberated when a nerve is stimulated electrically and that the fluid containing this substance will stimulate another heart without the intervention of nerve activity. The idea for this experiment occurred to him at 3 a.m. on two successive nights. The first night he wrote it down, but in the morning he found that his notes were illegible. The second night he took no chances, getting up and going to the laboratory immediately. He completed the crucial experiment by 5 a.m. Sir Henry Dale, who did much of his important work at the Wellcome research laboratories, showed that Loewi's *vagusstoff* was the small molecule called acetylcholine. They shared the Nobel prize in 1936. Two years later Loewi was arrested as a Jew but was allowed to leave Germany provided that he handed over his Nobel prize money to the Nazis.

It is interesting to ponder why nature employs a small molecule as a chemical nerve transmitter. The reason is probably that the 'firing' of a nerve is an 'all or nothing' effect: the signal passes or does not, with no variation in its intènsity. An obvious method of converting this digital response to a graded or analogue effect is to allow each nerve to emit a chemical at the end. Several nerves will each produce molecules whose total number is summed at receptors in the spinal cord before transmitting the signal to the brain. A finite number of these receptors will ensure that there is a maximum possible effect of nervous stimulation and will preclude overloading of the system.

The molecule which transmits the nervous signal across the various gaps in the nervous system is specific for each type of gap. All transmitter molecules are small, some with fewer than twenty atoms. Their effects are often dramatic, and the names of some of these transmitters have passed into common usage. Adrenalin is a small molecule which carries the message across the nerve gaps in conditions of 'flight or fright'. When an animal is startled or needs to defend itself, the heart beats faster because the signal causing this to happen has been received by the heart muscles; the crucial signal is carried by adrenalin molecules. In the brain there are billions of gaps between nerve cells. One of the transmitting molecules in the brain is called dopamine, a close molecular relative of adrenalin.

Receptors

It is believed that when a nerve transmitter molecule diffuses across one of the gaps in the nervous system, it binds to a 'receptor' in a cell across the gap, causing a change in this second cell. These receptors are thought to be protein molecules, with a binding site very specifically designed to receive the transmitter molecule, rather like the specific binding sites in enzymes, which are the proteins used in transforming molecules in chemical reactions. The original concept was that of a key fitting into a lock, but now it is clear that this would be an over-simplification. More probable is an arrangement where both the small transmitter molecule and its protein binding partner have some flexibility; a change in shape of the receptor molecule would then be the means of continuing the signal, triggering off further molecular changes inside the second cell.

Once the transmitter molecule has been bound and the signal passed on, it has to be released and removed; either it has to be reabsorbed by the transmitting nerve cell which originally released it, or destroyed by a suitable enzyme, leaving the original transmitting cell to produce fresh transmitters. Both systems are used in the body.

If the second mechanism is adopted, then the whole system is very vulnerable to anything which destroys or incapacitates the removal enzyme. This is, in fact, the destructive function of the nerve gases which were so much feared in the Second World War. Nerve gases attack the sites on the enzymes which bind nerve transmitter molecules prior to destroying them. Unable to remove the nerve transmitters, the whole nervous system will cease to function. It is akin to shutting down a factory not by destroying the machinery but by interfering with the control system.

The brain

The brain itself is made up of a vast number (about 100 billion) nerve cells. Each cell varies in length between 5 and 100 thousandths of a millimetre, and each is connected via synapses (junction gaps) to other nerve cells. A cell receiving several signals may integrate them and then emit an outgoing signal to another nerve cell. Again, within the cell, the transportation of the message is by electrical conduction, but chemical transmitters carry the signal across the gaps between cells.

Most of the nerve cells of the brain will receive a variety of inputs, some of which will excite them to send a signal and some of which will inhibit them. About thirty types of molecule have been postulated as chemical transmitters; some are excitatory and some inhibitory. These transmitters seem to be localized in specific clusters of nerve cells.

Although it is by no means proven, it does seem likely that some of the brain's most mysterious qualities—for example memory—are intimately related to the synapses which join cells. There could, say, be an increase in the efficiency of one synapse at the expense of others on the same cell. A particular combination of stimuli would then enhance one pathway among the many possible. Experiments on simple living systems have shown that when an animal learns or forgets a response, identifiable changes take place in the transmission of signals across particular junction gaps.

Of all the areas of illness, mental illness is least understood. However, it now seems more than ever likely that some mental diseases are molecular in origin. A slight imbalance in the production of the molecules which transmit nervous signals across junctions is manifest as a mental disorder. For example, brain nerve cells emitting acetylcholine are excitatory while those emitting dopamine are inhibitory. Parkinson's disease involves the progressive atrophy of nerve cells emitting dopamine, and an imbalance of acetylcholine results in the characteristic tremor, rigidity, and akinesia associated with the disease. If the imbalance is in the opposite direction, then it is suspected that schizophrenia is the result.

Drugs

If we can understand how nerve cells in the brain or elsewhere in the nervous system pass on their messages with small molecules, then there is an irresistible temptation for us to interfere with the system. By blocking or enhancing the effect of the transmitter molecules, we can perhaps cure disease with similar small molecules synthesized in the laboratory. This is one aspect of the work of the pharmaceutical industry, which we look at in the next chapter.

9

Drugs

Once the mechanism of biological control is understood at the molecular level, then it is possible for the chemist to interfere with the system to correct a diseased condition. In curing diseases of the nervous system, many of the products of the pharmaceutical industry work either by replacing the molecular nerve transmitter with another molecule which will do the same job (in pharmacological jargon, an *agonist*), or by blocking the receptor site with an alternative molecule (an *antagonist*).

Agonists

Agonist molecules are required to produce an effect similar to that of the body's natural nerve transmitter molecules and, in consequence, must bind to the appropriate receptor and also promote a change in shape or electron distribution similar to that produced by the natural molecules. Of course, one way of doing this would be to administer the natural transmitter molecules themselves, either directly or in a form such that the body's own chemistry could create extra transmitters.

This procedure is followed in the treatment of Parkinson's disease, which appears to be the result of an imbalance between the two transmitter molecules acetylcholine and dopamine. Administration of extra dopamine can alleviate the symptoms of the disease.

Often, however, administration of the natural transmitter molecule is not a helpful procedure because the body has enzyme mechanisms designed to remove extra transmitters. In this case, an agonist molecule that is capable of replacing the natural transmitter but is not removed or destroyed so quickly is preferable. The natural transmitter molecule and its synthetic agonist have to be interchangeable at the molecular level and, in consequence, able to produce the same physiological effect—perhaps contraction of blood vessels to the skin or contraction of the stomach, intestine, or radial muscle of the eye.

Antagonists

A far more widely used molecular device in the pharmaceutical industry is the antagonist: a molecule which will bind to the receptor for a natural transmitter, produce no effect, but prevent the natural transmitter molecule from performing its role.

The most widely used natural transmitter in the peripheral nervous system of the body is acetylcholine. The effect of acetylcholine at the junction between the nerve and the muscle may be blocked by an antagonist. The muscle may then be kept relaxed. Employed in surgery, such antagonists prevent muscles contracting when they are being cut or touched or when tubes are being fed down the alimentary canal of a patient.

Antagonists were discovered long before medicine became scientific by South American Indians, who used them to poison the tips of their arrows. The most famous example is curare, which is poisonous only when injected but not when eaten, presumably because it is broken down in digestion. It blocks the activity of acetylcholine. This is but one example of how modern science has learned much from the chemistry of primitive peoples.

In Western society, heart-disease is a major problem. Sufferers of irregular beating (cardiac arrhythmias) or angina have had their problems alleviated by blocking the effects on the heart of the nerve impulse transmitter noradrenalin, with the so-called 'ß-blockers', antagonists of noradrenalin (a compound very similar to adrenalin). As with the agonists, the antagonists have to be similar to the transmitter molecules if they are to replace them in the receptor sites. Usually they are slightly larger so as to allow better binding in the sites and hence to hold fast and exclude the natural molecules.

An antagonist drug is sometimes used to block the action of the transmitter histamine, which causes secretion of acid in the gut. Blocking of the effect allows stomach ulcers to heal. Yet again the drug molecule has not only to be similar to the molecule it replaces but also to have some extra parts which will allow strong binding to the macromolecular receptor.

Development of new drugs

The important antagonists mentioned—ß-blockers and histamine antagonists—were found by logical scientific research. Chemists started with a natural agonist and, with progressive changes in the molecular make-up, synthesized new molecules. But although the role of the chemist in drug development is that of synthesizing new molecules for testing, the skill lies less in the actual synthetic chemistry than in deciding which molecules to synthesize—because the number of possible variants of quite small molecules is enormous.

Not all new drugs are found in a totally logical way. Indeed, many of the most important therapeutic agents have been discovered quite by chance. It was alert observation which led to the use of penicillin as an antibacterial agent and benzodiazepines (such as Librium and Valium) as antianxiety agents. Random screening has also played a part, but the success rate is low. During the Second World War, some 14,000 compounds were tested as possible antimalarials but only a handful were selected for clinical trials.

The extent of the chemist's contribution to drug development can be judged from the fact that about a quarter of a million potential anticancer drugs have been synthesized and screened. The chemists tries to understand in molecular terms what kind of drug is required so that he or she can be more specific about the compounds synthesized and so avoid random screening.

Blocking enzymes

Therapeutically useful molecules are also designed to block the active sites on enzymes. An example is the drug used in cancer chemotherapy (particularly leukaemia) called methotrexate. A substance called folic acid plays a role in the production of cells, being converted by a particular enzyme (dihydrofolate reductase). Methotrexate blocks this reaction and has been a major contributor to the improved chances of children suffering from leukaemia.

Blocking the activity of an enzyme can also interfere with the effect of nerve transmitter molecules. If the enzyme which removes the transmitter molecules is blocked, then the number of these transmitter molecules will increase. The so-called monoamine-oxidase inhibitors (MAO inhibitors) raise the concentration of adrenalin by restricting the mechanism for its removal. This provides a mechanism for controlling blood-pressure but, clinically, molecules which block the enzyme are used as antidepressants. The only problem is that the same enzyme is used to break down other similar molecules (amines) which come from digesting food. Patients taking these enzyme blockers are occasionally subject to adverse reactions if their diets contain large amounts of particular amines or if they are taking other drugs which release amines which need to be removed. Thus foods like cheese or beef concentrates may have to be avoided.

Antibacterial drugs

Some of the most dramatic changes in human health have come from the employment of antibacterial drugs: sulphonamides and antibiotics such as penicillin. Their discovery goes back to the great German scientist, Erlich, who used the newly discovered synthetic dyes to stain cells and suggested that if some dyes stained cells of one sort but not another, then here was a potential source of drugs: chemicals which would bind to and kill cells of parasites or bacteria but be much less toxic to the host.

This idea was explored by another German Nobel prize-winner, Domagk, who worked for the BASF company. He showed the dye prontosil (a sulphonamide) to be an effective antibacterial drug in a most dramatic way—by using it to save the life of his daughter in 1936.

Of even more widespread significance was the demonstration by Florey that the compound penicillin, discovered in 1928 by Fleming but then ignored, was very potent against bacterial infections. The Second World War casualty figures when compared to the appalling statistics for the First World War bear testament to the efficacy of penicillin in reducing the number of cases of men who died from wounds.

Since 1945, chemists in pharmaceutical companies have investigated numerous substances like penicillin, which occurs naturally in a mould. Beyond this, they have investigated the chemical structure of these natural antibiotics and made synthetic variants. The chief reason for doing this is that many strains of bacteria eventually develop immunity to specific types of drugs.

Drugs for the healthy

A change in human life as equally profound as the use of drugs to combat illness has come from the introduction of drugs for the healthy. Above all, the contraceptive pill has probably done more for women's liberation than all the legislation in history. Controlling fertility has also set in motion political consequences which, as yet, we can only dimly recognize. In Western society some twenty years ago, the besetting problem seemed to be an exponentially expanding population; now schools are being closed as the school-age population falls.

All this has occurred as a result of our learning how the body produces hormone molecules and administering synthetic hormones to fool the woman's reproductive system into acting as if she were pregnant, and thus preventing conception.

The social effects

Medicine has been transformed since the Second World War by the use of drugs. Most of the drugs currently prescribed by physicians were not available a generation ago, and we are getting closer to the era of 'a pill for every ill'. Despite this, it is also clear that there is general public disquiet. Consumer demands for levels of safety are often strong enough to deny sufferers the benefits of a new drug for fear that, in a few rare cases, harm may result. The moral questions are pointed and serious. Above all, this problem was brought to the public eye by the thalidomide tragedy, the case of a drug used to prevent morning sickness during pregnancy that resulted in some babies being born with deformities.

The fact is that the living system is so complicated that it is not yet fully understood at a molecular level. In consequence, the unforeseen can happen—despite exhaustive testing on living animals, itself a distressing and morally questionable activity. But given the choice, say, of saving a child with leukaemia or sacrificing experimental animals, someone has to make a decision. Only by increasing our understanding of what is actually happening at the molecular level can these dilemmas be avoided. Until we are closer to the goal of complete understanding, all drugs should be taken with caution and with the realization that there are always dangers.

10

Food

People worry about taking drugs—quite rightly if this fear is because of the likelihood of side-effects. But such a fear is lunatic if it is based on the idea that eating chemicals is necessarily bad. All food consists of chemicals, of molecules. And a specific sort of molecule is identical whether extracted from nature or synthesized by a chemist.

We are, as we have seen, made up of molecules, some large and some small. The source of these molecules is our food. Some molecules (notably water) may be used by the body in the form in which they enter, but most are transformed by the enzyme molecules in the body to fulfil specific roles.

Cooking

One of the ways man shows his sophistication compared to other living creatures is by cooking his food before eating it. Cooking is used to modify molecules and thereby make them easier to digest. Digestion essentially consists of enzymes breaking down large molecules into smaller ones which can be dissolved and absorbed into the bloodstream. Cooking starts the digestive process off before the food is swallowed.

An essential component of diet is protein. Since so many of the body's structures are protein, a living system needs a balanced source of amino-acids to build its own proteins. Hence we eat other living things, and we can become ill if this balance of essential amino-acids is not provided. Animal protein such as meat will serve this need, and vegetarians who eat dairy products get their animal protein from milk or cheese. However, strict vegetarians who avoid all animal protein may need their diet supplemented with the component molecules that are otherwise too sparse in their food.

To make protein more digestible, we cook our food. Proteins, it will be recalled, are macromolecules made up from strings of amino-acids, rather like strings of beads. The linear chain is then folded and cross-linked to give a complex but very specific three-dimensional structure, unique for each protein and dependent upon the constituent amino-acid building blocks and their sequence in the chain. When heated, the three-dimensional structure of a protein becomes more random because the heat energy causes atoms and groups of atoms on the protein to shake and vibrate until the weak bonds which cross-link the chain are broken. Technically, this is called 'denaturing', with the loss of specific structure and the positions of the amino-acids randomized. It is the molecular equivalent of a wall of bricks collapsing. When the previously coiled and cross-linked protein chain is opened up in this way, it is much easier for the digestive enzymes which break down the chains to function. Part of the work has already been done.

One of the simplest forms of cooking is barbecuing, in which the meat is heated directly by a flame. The process of breaking down the three-dimensional protein structure occurs rapidly because a lot of heat energy is supplied. On the outside of the meat, which is hottest, charring occurs: the protein molecules react with oxygen, leaving a sooty deposit of carbon and other molecules rich in carbon.

A less drastic if slower form of cooking is boiling food in water, the maximum temperature of which is normally 100 °C. We can, of course, cook the food more rapidly if we use water which boils at a temperature higher than 100 °C. A higher boiling-point can be achieved in a pressure-cooker. Water boils when bubbles of steam rise from the bottom of its heated container and reach its surface. If the air pressure above the liquid's surface is less than normal atmospheric pressure, then the water boils at a temperature lower than 100 °C. This phenomenon is well-known to mountaineers, since the drop in pressure at high altitudes is significant: on a 20,000-foot mountain, it may take 15 minutes to cook an egg because of the lower temperature of boiling water. By contrast, in a pressure cooker, the air pressure above the water is increased by sealing the steam in the container. At the higher pressure, the boiling water has a temperature well above 100 °C, and therefore more heat is provided to break down proteins or starch in food.

Alternatively, we can heat the food in a liquid which has a higher boiling-point than water. Oil or melted fat, used in frying, has a boiling-point around 200 °C.

The heat supplied to food in cooking is taken up by the food molecules in the form of increased molecular vibration and rotation. The heat from a fire or oven contains a wide range of energies—from long-wavelength energies, which cause molecules to move faster, to intermediate or microwave energies, which cause small molecules to rotate faster, and to infra-red radiation, which causes molecules to vibrate.

By contrast, the microwave oven is a controlled device which produces all its energy in one narrow energy range. On the one hand, it makes water molecules rotate faster and is thus an ideal way of heating frozen foods very rapidly. On the other hand, since the cooking dishes do not contain loosely held molecules capable of rotating, they are not warmed in a microwave oven. Hence we get the seemingly bizarre effect of hot food on cold plates.

Tenderizing

A refinement of cooking is to use enzymes to break down some of the protein structure before eating. This trick was known to cooks for centuries before its molecular basis became clear. The most obvious application of the idea is pineapple with gammon: fruits such as pineapple are particularly rich in enzymes which break down proteins, and mere contact is enough to soften or tenderize the meat and render it more digestible.

Commercially, this idea has been exploited with the use of enzyme preparations to tenderize steaks before they are cooked. The meat tastes particularly tender because it has been partially digested before it gets on to our forks!

Vitamins

Vitamins are substances which are essential for a healthy life and which the body is unable to synthesize from other molecules. Man is unusual in being required to eat many more vitamins than do other creatures. It seems likely that, in the process of evolution, as we gained new qualities by chance mutations of our DNA, we lost some instructions in the code which were the blueprint for the synthesis of vitamins.

Because only minute quantities of vitamins are required, the process of finding out what a vitamin is and how much of it is 'essential' has been complicated. Only by studying the effects of abnormal diets can conclusions be clearly drawn. Our reliance on vitamin C, for example, was shown when a deficiency of it, resulting from a lack of fresh fruit, led to scurvy in sailors (causing British sailors to eat limes—hence their American nickname 'limey'). Animals other than man are not suitable models since their genetic molecular make-up differs from ours; for example, rats, chickens, and dogs can synthesize vitamin C. Thus experimental work is hampered by the ethical problems of working on human beings. Some of the best and clearest studies of vitamin deficiency problems were performed on volunteer conscientious objectors during the Second World War.

The classification of vitamins is the commonly used alphabetical one. Vitamin A has an important role in vision. Light falling on the molecules alters their structure, and this structural change is transmitted to the brain. The vitamin B group contains a number of different molecules which are essential for the activity of several crucial enzymes. Vitamin C, although it is the simplest vitamin, is the centre of considerable controversy as to whether massive doses relieve the common cold or even cure cancer. Conservative scientists are sceptical. The experience of advocates of vitamin C prophylaxis such as Linus Pauling, another two times winner of the Nobel prize, seems to be that over the years more and more vitamin has been required for protection. Initially, milligrams sufficed; now grams are needed. Could it be that the early vitamin, extracted from natural sources, had impurities which were the real active ingredients? Since the fad has grown, the product has been produced in purer and purer form, and it is now totally synthetic and very pure indeed. If the impurity were the active principle, then this would explain the apparent need for increased doses. Vitamins of the D group are molecules of the steroid type, related to cortisone and the sex hormones, while vitamin E is an antioxidant and vitamin K is essential for prothrombin production in the liver.

Flavours

The taste of food is again controlled molecularly and, in consequence, is open to the chemist for manipulation. The same type of mechanism used by enzymes to bind to molecules or by receptors to interact with nerve transmitters is employed: tasting a particular flavour is the fitting of a small molecule into a site designed to accept the shape and electron distribution of that precise molecule or of one very similar. Hence, for a chemist, the story of producing artificial flavours runs parallel to that of producing drugs. He or she has to synthesize molecules which have the same or similar shapes and electron distributions as the natural flavour molecules so as to fit the taste receptor molecules of the body.

The first appearance of the chemist in the story concerns the isolation of natural substances which add flavour (for example the taste of strawberry). The structure of the pure molecules is then determined, and ultimately the chemist tests variants which he or she has synthesized, often resulting in much stronger flavours.

The search for alternatives to sugar (for sweetening without fattening) created an industry based on saccharin, but this product is not a good substitute because of its after-taste. Industrial chemists have now produced an alternative with only a couple of amino-acids joined together. This new product (known to chemists as aspartame) is many times sweeter than sugar and can be metabolized and excreted by the body's existing enzyme systems, which are designed to split bonds between amino-acids. There are virtually no worries about the possible side-effects of the new sweetener. Soon all our cola will contain it.

Preservation

The public demands not only food with flavour but, of even more importance, food that is not poisoned by bacterial toxins. The chemist's answer to this problem has been to use the same strategy he or she uses in chemotherapy: the development of compounds which exhibit selective toxicity, lethal to the intruder but benign to the host. This is a principle applied to a wide spectrum of problems, including those connected with agro-chemical research as well as, more specifically, the question of what preservatives to use in food products. It is important to synthesize, and to confirm the efficacy of, molecules which will kill a pest but, at the same time, will not harm human beings. Occasionally the results are surprising. Preservatives are added to bread to delay staleness. The use of these additives was opposed by a strong pure food lobby, particularly in the USA. Not only has it proved safe but it has also resulted in a dramatic reduction in stomach cancers, probably because the compounds added to stop the bread going stale also remove some cancerous agent found in the stomach.

11

The chemistry of energy

At one level, food is a source of the molecules from which the body synthesizes new and essential products. At another level, food is also our source of energy.

Molecules and energy

Slimmers are familiar with 'calories': measures of the amount of heat released when burning standard quantities of substances. To measure the calorific value of a substance, a fixed quantity is burned in oxygen to give mostly carbon dioxide and water, since organic molecules contain mostly carbon and hydrogen, and the amount of heat generated is measured. Hydrogen gas is the most efficient fuel in terms of the heat emitted per gram of fuel burned. Gasoline is less efficient by a factor of three, but it is relatively easy to distil from crude oil and it is convenient to transport. For animals, the main fuel molecules are fats, which yield almost as much energy per gram when burned as gasoline.

By contrast, plants use as their main fuel molecule the long-chain sugar polymer starch, which is almost three times less efficient in calorific terms than fat. The reason why evolution has resulted in this solution to the problems of energy storage is probably because the chemistry involved in extracting energy from starch is much simpler and quicker than is the case with fat. Unlike plants, animals need to be mobile, and fat is much lighter than starch, even though the chemistry of energy extraction is more complex. Animals do need, however, the rapid energy access of starch fuel and consequently have glycogen (a starch derivative) in their bloodstreams.

The amount of energy which can, in principle, be extracted from a chemical reaction is found by measuring the difference between the energy liberated when burning the reactant molecules and that liberated when burning the products of the reaction. The amount of energy released in a chemical reaction depends on just how comfortable the electrons are in the alternative molecules constructed from the same atomic building blocks. The tighter the electrons are held after combustion, the more stable the product molecules are and the more energy is produced as heat.

At the molecular level, the heats of combustion can be interpreted in terms of the energy required to break particular chemical bonds in a molecule. With hydrogen and oxygen, the bond energies are together far more than the energy required to break an O–H bond in H_2O; therefore hydrogen and oxygen react explosively to give water and a great deal of energy.

Batteries

One of the most satisfactory ways of extracting energy from chemical reactions is as electricity rather than heat. This is done in batteries.

The chemical reactions which form the basis of batteries depend on the loss or gain of electrons from atoms or molecules. If the donor and the recipient of electrons can be isolated, then electrons will pass from one to the other and there will be a flow of electric current through an external wire or circuit. For example, metallic zinc has less affinity for its electrons than does copper. Thus if we have zinc in a solution of zinc ions which have lost two electrons and are thus doubly positively charged (Zn^{++}), connected to copper in a solution of similarly dipositive copper ions, (Cu^{++}), we can create a flow of electrons (a current) from the zinc to the copper in a wire joining the two metals: a simple electrical cell.

The batteries used in bicycle lamps, portable radios, and tape recorders are a variation on this principle and are known as dry cells. One pole of a dry cell (the positive anode) is a central rod of carbon with a metal terminal, while the other (negative cathode) is the zinc which forms the outer case. The substance between the two electrodes (called the electrolyte) is not a solution but a paste of zinc chloride, ammonium chloride, and manganese dioxide. At the zinc electrode, zinc atoms become zinc ions, yielding up two electrons, while at the anode, the manganese and the ammonium ions each gain an electron. Dry cells deliver about 1.5 volts of power, but since ammonia gas is produced round their carbon rods, the batteries run down. Left for a while, they recover somewhat, as the ammonia combines with the zinc ions. The cell cannot, however, be recharged, as can the lead accumulator used as a car battery—a dynamo drives a current through the battery and restores the chemicals to their original states.

Because of the energy crisis and the desire to produce non-polluting transportation, many chemists are involved in trying to devise better, cheaper, and lighter electrochemical sources of electricity, including the use of solar energy. This latter approach is yet another example of the chemical way of thought: seeing how nature does things and then reproducing this mechanism, with the ultimate aim of improving on nature.

Energy transformation

In nature, there are two supremely important mechanisms of energy flow: respiration, which involves the breakdown of glucose to yield useful energy; and photosynthesis, which is the trapping of the energy from the sun to synthesize glucose. While both plants and animals burn their foods with oxygen to produce stored energy as a molecular fuel, carbon dioxide, and water, only plants can use the sun's energy to recombine water and carbon dioxide into sugar. Thus, ultimately, all our energy comes from the sun, since animals live off plants. Even the fossil fuels, oil and coal, are the remnants of living systems transformed by geological influences.

The mechanism of glucose metabolism is common in most respects to all living things, yet again illustrating the unity of understanding afforded by the chemist's view of life at the level of the molecule. Both respiration and photosynthesis are much studied by chemists, with the scientific aim of understanding how they work. Typically, there are chemists who want to interfere with and block the processes—for example, by the synthesis of herbicides, such as paraquat, that block photosynthesis. Others study photosynthesis in the hope of coming up with devices which will perform the same function either better or more conveniently. This may seem a very arrogant attitude, contemporary scientists bettering something which nature has taken millions of years to evolve. In fact, however, nature has had to produce its machinery using a limited range of materials which are stable over only a narrow range of temperatures. In consequence, it is not a forlorn hope that we may be able to do better than nature. Photocells, which convert light into electricity and are used to drive satellites, represent an efficient, if rather expensive, example.

The chemist's contribution to solving the energy crisis

Although public attitudes towards the energy crisis wax and wane with the price of petrol, scientists are constantly aware of the challenge facing humanity.

Crude oil and coal were formed from the remains of small marine animals and plants which were buried on the sea-bed millions of years ago. The constituent molecules are mostly hydrocarbons—that is, molecules containing only carbon and hydrogen atoms. In the oil industry, this mixture of molecules is separated by distillation to heavier molecules (used in lubricants), lighter molecules (used in gasoline), and very small molecules (which form natural gases). But these two resources are finite. As far as we know, no oil or coal is being formed in the earth's crust today.

The contributions of chemistry to finding other energy sources span most of the fashionable remedies. Many advocate using hydrogen as a fuel. To this end, researchers are trying to harness the sun's energy in order to split water into hydrogen and oxygen and to devise materials in which hydrogen can be stored, perhaps by absorption into a solid.

Electrochemical possibilities include not only new types of battery but also solar energy converters, seen by many as the most acceptable long-term solution. In the more immediate future, applied research is seeking to improve coal as a source not only of energy but also of the small molecular building blocks or monomers for the plastics and artificial fibre industries.

The chemist's contribution to the nuclear option is twofold. With nuclear fusion, it is largely restricted to developing new materials. With nuclear fission, it involves nuclear chemistry: the transformation of atomic nuclei into other nuclei, with a consequent release of energy. Options such as fast breeder reactors clearly depend on the work of nuclear chemists, as, of course, do all the safety measures imperative in the nuclear industry.

By and large, the problem of energy is one of physics rather than chemistry, but the materials themselves fall within the chemist's bailiwick.

12

Giant molecules—man-made materials

Until the mid–1920s, few chemists believed that a stable molecule containing several thousand atoms could exist. The common belief was that the insoluble substances isolated from natural sources were merely aggregates of small molecules held together by some mysterious force. It was in 1922 that the German chemist Staudinger proposed that substances such as rubber were simply large molecules (he used the expression 'macromolecules') made up out of a series of small molecular units, rather like beads on a string. In the jargon of this branch of chemistry, giant molecules or *polymers* are made up from *monomer* sub-units (individual small molecules).

Natural polymers such as rubber, wool, and cotton have been known since ancient times. As we have seen, the macromolecules of living systems—proteins, nucleic acids (DNA), and carbohydrates—come into this category, with the monomer building blocks being, respectively, amino-acids, bases, and sugar molecules. Following the insight of Staudinger, for which he was awarded the Nobel prize, there was a rapid development by chemists of man-made or synthetic polymers, giving rise to the plastics and artificial fibre industries.

Few innovations have had more effect on daily life than polymers, encompassing nylon stockings, drip-dry shirts, polythene packaging, toys, sports equipment, and almost every aspect of day-to-day existence. By contrast, the initial developments were prosaic: synthetic polymers were first produced as electrical insulators.

Synthetic polymers

The first synthetic or man-made polymer was produced in the USA in 1907 by Bakeland, a Belgian-born American industrial chemist, who was searching for a way to make artificial shellac (natural shellac is a resin excreted by an insect, the lac bug). This is a typical chemical intervention: trying to replace an expensive natural product by a laboratory-made alternative. Bakeland heated under pressure a mixture of two organic substances, phenol and formaldehyde—respectively used as an antiseptic (carbolic acid) and as a preservative. Upon cooling the result, he obtained a clear, hard, amber-like material in the shape of the heating vessel. The material, now known as bakelite, is an excellent insulator and is used to make electrical plugs and switches.

Polymers like bakelite are called thermosetting plastics. The basic resin—in this case, a gummy mess of phenol and formaldehyde—is put into a mould and then heated. This results in chemical bonds forming between the polymer chains (composed of alternating phenol and formaldehyde molecules) and gives rise to the rigidity. Thermosetting plastics, once formed, cannot be melted and remoulded. Indeed, if they are heated to high temperatures, they decompose—giving the characteristic burning smell of over-heated plugs and electrical gadgets.

Polymers which can be melted, poured into moulds, and remelted are, by contrast, called thermoplastic. An example is polyethylene (polythene), produced in Britain by ICI. Polyethylene was found largely by accident when ICI chemists were studying reactions of gases at high pressures. The small gaseous ethylene molecule can be induced to take part in a chain reaction in which the string of added molecules just gets longer and longer, until the resultant macromolecule is so big that the substance forms a white solid. This polymer is thermoplastic, so called because it can be remelted and remoulded indefinitely. Again, polyethylene is an excellent electrical insulator and played an important part in the Second World War as an insulator in radar apparatus. In peacetime, it has become common not only as electrical insulation but also in a wide variety of applications, especially moulded kitchen-ware.

Polymerization

As we have seen earlier, molecules are 'happy', or more accurately 'stable', if all the constituent atoms have the magic number of electrons orbiting around them. This may be achieved either by losing or gaining electrons to form positive or negative ions or, more commonly, by sharing electrons to form covalent bonds. The monomer reactants of polymer chemistry are generally single stable compounds which conform with notions of stability: their electrons are all paired. To achieve polymerization, the reaction has to be initiated by an unstable molecule, often one with an odd number of electrons—or a 'free radical' as chemists would say. The free radical combines with a reasonably stable monomer molecule to give another free radical with an unpaired electron. This, in turn, combines with another stable monomer to give yet another free radical, and so on, in a chain reaction, until the chain is broken by two odd-electron radicals combining to give a stable product. Polymerization chain reactions thus involve: initiation by a free radical; propagation; and, finally, termination. If a stable molecule is split into two radicals, then we can also get branching of the chain.

Chemists have played many variations on this theme and, in so doing, adjusted the properties of the resulting polymers. Commonly encountered examples of this procedure include: polyvinyl chloride (PVC), used in gramophone records or floor coverings; polyvinyl acetate, used in chewing-gum and adhesives; polystyrene, used in moulded objects and kitchen-ware; and polyperfluoroethylene or Teflon, used as heat-resistant non-stick coatings for cooking utensils. Particularly important historically was the synthesis of artificial rubber, without which Germany could not have embarked on the Second World War, since it had no access to supplies of natural rubber. The USA also made a gigantic effort to produce artificial rubber when the Japanese occupied the rubber plantations of the Far East.

Most of the polymers discussed so far have simple carbon-containing molecules as monomers, but a new dimension was added when silicon-containing monomers were introduced. Polymers containing silicon rather than carbon atoms (i.e. silicones) are valued for their heat-resistant properties. If some of the silicon atoms are replaced by boron atoms, strange substances like 'potty putty' can be produced. The boron atoms readily form cross-links between chains in the polymer by accepting electrons from oxygen atoms to form chemical bonds. When a sample of this 'potty putty' is pulled steadily, it extends like a piece of Plasticine. At the atomic level, the silicone chains of alternating silicon and oxygen atoms slide over each other as the boron atoms form bonds successively with each passing oxygen-atom neighbour. When pulled sharply, the material breaks like a piece of cheese because the boron bonds are unable to progress smoothly from one oxygen atom to the next in the chain. The substance can thus be plastic or brittle depending on the conditions under which it is handled.

Many of the substitutes for cotton and wool differ from the simplest polymers in that two different molecules (A and B) alternate in the chain—as in bakelite. Nylon and Terylene are examples: instead of having a repeating monomer molecule of type A, giving –A–A–A–A–, the structure is –A–B–A–B–A–B–.

The properties of the polymer depend not only on the nature of the repeating units in the polymer chain but also on the stereochemistry or spatial arrangements of the atoms. In the early 1960s, a significant step in determining the stereochemical aspect of polymerization was achieved by Ziegler and Natta. They introduced catalysts, in the form of organometallic compounds, to control the polymerization reactions. These compounds have simple organic, or carbon-containing, groups joined to a metal atom and produce polymers with a predictable geometry—similar to the role played in nature by the enzymes which control the synthesis of biopolymers such as protein and DNA. For the chemist, aping nature is a constant goal, and few of nature's devices have more potential than enzymes.

Economics of the polymer industry

The polymer industry boomed from the 1950s up until the oil crisis of the mid-1970s. Its feedstocks, or sources of monomer, were the hydrocarbon molecules found in oil and natural gas. When oil was cheap, man-made materials were less expensive than their natural rivals like wool or cotton. The oil crisis changed this picture, perhaps for ever. It now makes little economic sense to build petrochemical plants in the developed world. It is far cheaper to create the polymers in oil-producing countries such as Saudi Arabia and so avoid having to transport the oil.

Furthermore, the cost of producing a new type of plant for a novel polymer is so great—the 'entry fee' runs into hundreds of millions of pounds—that it is no longer wise to embark on new products. A whole industry and area of chemical research has thus moved in one single adult lifetime from being novel, exciting, and attractive to being depressed, contracting, and old.

Perhaps nowhere else is the close contact between chemistry and 'real life' more apparent than in the polymer industry. Polymer chemistry deals with everyday products and hence with major industry. All plastic products are molecular when viewed at the microscopic level. (Little wonder that chemists tend philosophically to be reductionists!) Because of this, the constraints of the world and corporate finance influence what is done in the laboratory and in the factory far more than they do in other academic disciplines.

For the future, existing polymer products will have to compete with natural rivals. Scientific advances are likely to be restricted to speciality products with specific and desirable properties such as flame resistance or electrical conductivity. Alternatively, a whole variation on the industry could start with coal as the feedstock rather than oil. Any development will of course have to be justified on the basis that products can be made more cheaply than natural or existing alternatives.

13

Chemistry for pleasure

Most of the developments of chemical research discussed so far have been unashamedly practical. Chemistry as a discipline has a long and honourable record of improving the material aspects of life. But it has also used the manipulation of molecules to provide pleasure. Our pleasures are governed by our senses and, in man, the most developed is that of sight.

The chemistry of colour

From far back in history, bright colours have been found to be attractive and to give pleasure—in their least sophisticated form as the bright sparks of fireworks. We have seen (in Chapter 1) how atoms and molecules are stable when certain numbers of electrons orbit their atomic nuclei. If sufficient energy is provided to a molecule, bonds may be broken and molecules transformed into simpler ones or even separated into unstable atoms. When the energy provision is more limited, the molecules can absorb the energy without being destroyed—by adjusting the orbits of the electrons and making them circle further from the atomic nuclei. The molecule or atom is then said to be in an 'excited electronic state'. These excited states are not stable, and the excited electrons may revert to their original orbits, at the same time emitting their previous excess of energy as light. The emitted light may be observed as an emission spectrum: the light is split into its constituent wavelengths using a prism, just as sunlight can be shown to consist of a 'rainbow' of colours.

With fireworks, chemical reactions provide the energy to generate the excited electronic states and subsequent coloured emissions. Different substances provide the range of colour. A simple illustration is provided by sprinkling salt (sodium chloride) into a gas flame, giving rise to the bright orange-yellow sodium emission spectrum, the same colour observed in yellow street lighting.

All coloured substances appear so to us because of the inverse of the mechanism of the emission of light. When white light, which contains all the visible wavelengths in sunlight, falls on a blue carpet, the carpet appears to be blue because all the colours except blue are absorbed, whereas blue is reflected. In such circumstances, the red or green wavelengths of light excite molecules in the carpet, but instead of being re-emitted, the energy is changed into small amounts of heat.

Until the nineteenth century, the dyes used to colour clothing were all naturally occurring substances. Some, such as madder and indigo, were derived from plants; others, such as lac and cochineal, had insect origins. There were readily available sources for red, blue, and yellow, but the difficult one was purple, derived from the shellfish murex by the Phoenicians and used to distinguish rank or aristocracy. Roman senators had purple stripes on their togas. In the nineteenth century, purple was obtained from lichens scraped off stones on the seashore.

Such crude process and the consequent scarcity and expense of dyes was changed dramatically when the chemist intervened.

Synthetic dyes

The synthetic dye industry dates back to the British chemist Perkin in the nineteenth century. At the age of 18, he was trying to make quinine. His experiment failed, but as he was about to pour away the mess, he noticed a purplish glint in the material. He added alcohol, and dissolved out aniline purple, a substance found to be capable of dyeing silk. In France, Perkin's discovery was followed up and gave rise to what was known in fashion as 'the mauve decade'. In England, industry moved more slowly, although Queen Victoria wore a dress dyed in 'mauveine' for the Great Exhibition of 1851 and, in 1862, postage stamps were coloured mauve.

Perkin's method was a one-off synthesis of a dye. A more general method was discovered by the German chemist Griess while working in a brewery in Stoke-on-Trent. He, like Perkin, started with the simple chemical aniline (obtainable from coal tar); but he found a way of converting it to a related compound, called a diazonium salt, which could couple with a second molecule to give a dye whose colour depended on the choice of the second molecule.

Griess's discovery was exploited in Germany, to which he returned. German industry produced diazo dye molecules, including Bismark Brown and alizarin. The importance of sulphonation was discovered by Caro at the German firm BASF, while the German organic chemist Baeyer found out how to synthesize indigo, thereby ruining the Indian indigo trade. By 1914, Germany so dominated the dyestuff business that khaki dye for the uniforms of the British Expeditionary Force could be obtained only from Germany, and therefore some of the troops had to be sent to the front in blue uniforms dyed in the natural woad of the ancient Britons.

Modern developments in dyestuffs have made available fast, non-fading, bright colours to fashion designers at a cost which does not preclude bright apparel for the less wealthy. But ironically, in the almost universal current fashion for denim jeans, the desired faded look is best achieved with the old, original, natural indigo!

Colours are used not only in dyes for clothing but also in paints. Here the chemist has provided not only a wide range of colours but also novelties in the nature of paint itself. The introduction of plastic emulsion paints and non-drip paints has revolutionized the world of home decorating. The result is that homes are very much brighter and more visually attractive than they were a generation ago.

Dyes with special functions are vital to the performance of colour film. Here chemists have developed molecules that are sensitive to specific wavelengths of light.

One advertising claim which is scientifically genuine is the addition of 'brightness to whiteness' in washing-powders. This seemingly impossible trick is achieved by including, in the detergent powder, substances which will absorb ultraviolet light and then re-emit some of this light in the visible region. A white shirt will then reflect not only all the visible light which falls on it but also some of the ultraviolet light; and as well as being whiter than white, it will also glow in the ultraviolet light commonly found as illumination in discotheques.

Molecules and smell

Nature uses as its mechanism for detecting scents the same device we have already encountered several times: the fitting of a small molecule into a specifically tailored slot which reflects its shape; the flexible key in the flexible lock. This recognition contrivance, as we have seen, is the functional basis for transmitting nerve impulses and for detecting flavours.

As in so many other chemical applications, the chemist's role in creating scents has been to try to ape nature. The perfume industry originally used some very exotic scents such as muscone and civetone, obtained from the scent glands of animals. When chemists managed to work out the exact nature of the active scent ingredients and synthesize replicas of the natural products, the way was open for a synthetic perfume industry. Perfumes can now be made more cheaply and are no longer accessible only to the rich.

For human beings, smell is not our most highly developed sense, but for many other creatures it is the chief source of information. So highly developed is the recognition of specific molecules by insects that they can sometimes detect single molecules such as pheromones.

The wilder shores of pleasure

Not only can the chemist's skills produce molecules which please our visual and olfactory senses but the same skills can be used in the far more dangerous area of fooling the brain in its appreciation of stimuli.

Since time immemorial, man has used alcohol to heighten sensation and to befuddle the brain. Other more lethal natural products such as opium and cocaine also have a long history. Once these products could be extracted in pure form and in large quantities, a blacker side of chemistry appeared. The chemist used his or her skills to produce narcotics for the underworld. Beyond that, new and even more devastating drugs of abuse, such as LSD, have been created.

Now, as the role of small molecules in the functioning of the brain starts to become unravelled, prospects both exciting and terrifying begin to emerge. Interfering with the transmission of impulses across particular junctions in the central nervous system may produce feelings of ecstasy or terror and often result in severe brain damage. The power to control men's minds will come with the knowledge of how they work. Given the record so far of the exploitation of narcotic drugs by unscrupulous operators, the prospects for the future must be viewed with grave concern.

14

Chemistry for profit

Most of the largest companies in the world are producers of chemicals. Among these are the oil companies, although arguably they are chemical in a rather restricted sense because they extract the oil (a mixture of hydrocarbons) directly from the earth's crust; they do not manufacture it. However, quite sophisticated chemical techniques are involved in separating the oil into different hydrocarbon fractions (bitumen, fuel oil, kerosine, gasoline, natural gas, etc.) and in converting one fraction into another. Most major oil companies also have associated activities which are chemical in a more obvious sense. Starting (usually) with natural gas, they synthesize and market chemicals which are raw materials for, among others, the plastics and agrochemical industries. Nowadays, the super-giant oil companies are very actively moving into these other areas as a precaution against the time when the fossil fuels run out or become uneconomic.

Not far behind in terms of size are the true chemical companies, vast, multinational, and, in general, very profitable. They are new industries—most were created during the past hundred years—with enviable records of profit, industrial relations, and safety. As Figure 14 shows, world chemical production has grown far faster than the gross world product during this century.

The table on page 83 lists the chemical innovations that have had a major industrial impact since the mid-nineteenth century. It is noteworthy just how many big industries had their origins within the lifetimes of today's employees. The plastics and synthetic fibre industries, for example, only started in the 1930s and those making pharmaceuticals and plant protection products (herbicides and pesticides) in the 1940s. None the less, the products of these industries are now to be found throughout the world.

Major chemical innovations since the mid-nineteenth century

1840	Gun-cotton Chloroform used as an anaesthetic First synthetic dye
1860	Nitro-glycerine Solvay soda ash process Celluloid Contact process for sulphuric acid Brin's oxygen process
1880	Phenacetin Castner brine electrolysis Phosphorus production Aspirin Acetylene Liquefaction of air Halocaine Viscous thread
1900	Bakelite Haber process for ammonia Ethylene production Nitric acid from ammonia Detergents Phenol-formaldehyde resins Synthetic rubber
1920	Rayon Cellulose lacquers Polystyrene Perspex Prontosil sulphonamides Nylon Polyvinyl chloride (PVC) Penicillin
1940	Polyethylene Polyurethanes DDT Silicones Epoxy resins Selective weed-killers Pure silicon Polypropylene
1960	Nuclear magnetic resonance Laser applications Biotechnology Inorganic materials
1980	

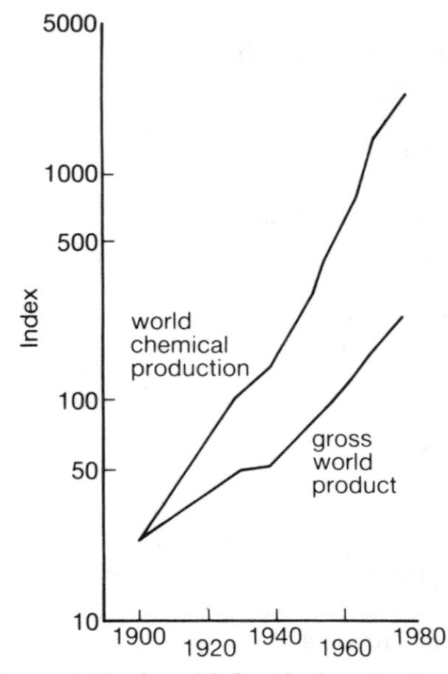

Figure 14. The growth of world chemical production since 1900 compared with the rise in the gross world product

Heavy chemicals

It is often stated that the economic and technical development of a country can be measured by its production of sulphuric acid. This is because the acid is necessary for the production of such basic materials as fertilizers, paints, fibres, plastics, dyes, and steel. Storage problems require, in addition, that production must respond rapidly to any change in demand. Sulphuric acid is an example of a 'heavy chemical'; one used in very large quantities. In 1980 annual world production exceeded 100 million tonnes.

Another example of a traditional heavy chemical is sodium hydroxide, or 'caustic soda', which is obtained by the electrolysis of brine, a process which also produces chlorine gas. Manufacture of heavy chemicals tends to occur in regions where the necessary raw materials are available. Burning limestone (calcium carbonate) to give lime (calcium oxide) is an ancient industry. Lime is 'slaked' with water to give the cheap alkali calcium hydroxide, which is again a major industrial raw material. In the eighteenth century, it was discovered that slaked lime absorbs chlorine gas to yield a convenient solid product which, when mixed with water, acts as a bleach that is again widely used in industries such as textiles.

The heavy chemicals industry now also enhances more modern products, which are used in huge quantities by other manufacturing industries. Many halogen-containing organic molecules are important as solvents and are used in huge amounts. Similarly, as discussed below, the monomers for the plastics industry have to be produced in very large quantities.

Petrochemicals, plastics, and fibres

In the nineteenth century, the chemical industry's raw materials for the manufacture of organic products were coal, salt, minerals, molasses, and fats. In the second half of this century petroleum became the cheapest and most convenient feedstock. The petrochemical industry takes hydrocarbons from the oil industry or from its own resources and breaks down the molecules into small molecular units; these units (or monomers) are then polymerized to produce plastics, as discussed earlier. Some plastics are extruded into fibres for the artificial fibre industries.

The chemical industry has become adept at adjusting the properties of synthetic polymers to provide precisely the qualities required for a specific product. One example concerns the texturing of yarn. Continuous filament yarn resembles an array of infinitely long rods and is thus limited to producing thin, smooth fabrics. In the mid-1950s, however, techniques were developed to process the filaments into helical or saw-toothed shapes so that a bundle of these modified filaments would form a voluminous low-density yarn resembling wool. This texturing process has led to such major innovations in fashion as nylon stretch pantihose.

The goal which the artificial fibre industry still sets itself is to produce fabric without the intermediate stage of spinning a one-dimensional yarn spun from fibres. There has been progress in producing non-woven fabric by heating two different types of fibre placed side by side so that they are welded together. The two different component fibres may have different desirable qualities, one, for example, conferring strength and the other dyeability.

Dyes and other fine chemicals

Fine chemicals are produced in relatively small quantities, usually because only small amounts are required to produce their specific effects. There is a continual demand for dyes which can be applied to fibres more efficiently by processes which use less energy and which produce less coloured effluent. The fastest dyes, which are held 'fast' and do not fade, are those where the dye molecule actually reacts and forms a strong covalent chemical bond with the molecules of the fibres.

Dyes are also incorporated into paints. Paint itself is not a pure substance but a mixture of components in solution and suspension. Here, the problem of innovation is to manufacture a product which as far as the user is concerned behaves in a homogeneous way even after storage for long periods and at the same time produces a resistant, dry, thin film which will cover and protect for many years. As is typical of the world of chemistry, man-made polymers are leading to the replacement of linseed oil formulations. The particular synthetic polymers used are designed to meet the needs of customers such as the automotive industry and domestic uses of emulsion paints. In particular, the demand is for polymers which are soluble in water. A polymer which is soluble in water as well as in an organic oil will make the cleaning of paintbrushes under the tap much easier; it will also find an application as a dispersant of oil slicks.

Pharmaceuticals

The most profitable parts of major chemical companies are those involved in health care, where the 'added value' of products is highest. Of all areas of chemistry this is also the area where the proportion of revenue spent on research is highest. Sums of $100 million are spent on specific molecular products before any of the compound is tried in a clinic. The rewards for success are equally impressive. A major drug may create perhaps a billion dollars of business in a year. But of course the chances of success for any particular compound synthesized are very slim, of the order of 1 in 15,000.

The difficulty of ensuring a financial return on research, particularly because of the risks of unwanted side-effects that could result in astronomical legal damages, is having a pronounced influence on the type of research conducted by the industry.

Among the major target areas of most of the leading drug companies are diseases of the cardio-vascular function—heart disease; arthritis and inflammation; diseases of the central nervous system; peptic ulcers; and the control of fertility. These are all areas where the sales potential can justify the expenditure on research.

In addition to discovering compounds which are useful therapeutically research must also be conducted into the formulation of drugs into a medicine. Drugs must be efficient in how they release their contents after administration. The releasing properties of tablets will be influenced by the cohesion and friction of their constituent powders, and the size or water content of their particles. A major research activity is the testing of polymeric materials as membranes or biodegradable coatings. If drugs are in these plastic coverings then perhaps close control may be exercised over the speed of release.

The pharmaceutical industry manufactures products not only for human health care but also for animal health and animal husbandry. There are drugs to control parasitic infections such as worms or liver fluke, while synthetic analogues of prostaglandins may be used to control the breeding cycle of domestic animals, leading to more efficient farming and cheaper food production.

Agrochemicals

At the heavy chemical end of the agrochemical range are the nitrogen fertilizers. Possibly the most far-reaching scientific research of this century was that carried out by Haber, who produced ammonia from nitrogen and hydrogen using high pressures and a catalyst. Ammonia may be oxidized to nitric acid, itself an important industrial heavy chemical. Fertilizers are usually nitrates or ammonium salts—both types of molecule contain the nitrogen atoms essential for the molecules involved in plant growth and health. Without the nitrogen fertilizer resulting from the Haber process, millions of people would be starving at the present time, over and above those starving in the arid parts of the world.

The chemical industry's influence on agriculture extends beyond the production of fertilizers. Herbicides, fungicides, insecticides, and plant-growth regulators are all used to improve yields and to make farming more efficient. This aspect of the agrochemical business has many affinities with the pharmaceutical industry: relatively small quantities are used and it is essential that any products are safe to use and do not present a hazard either to consumers of the treated crop or to the environment. Similar also is the need to formulate products that can be administered both with safety and with the maximum biological effect.

Chemical engineering

The Haber process mentioned above is a good illustration of how the engineering aspects of chemistry are vitally important when scaling up a piece of laboratory chemistry to the industrial scale. Haber worked with a relatively simple piece of apparatus which stood on a laboratory bench. It took him three years of experimenting to define the temperature and precise conditions of temperature and pressure which enable nitrogen, N_2, and hydrogen, H_2, to combine to produce ammonia, NH_3, using a metal catalyst to speed up the reaction. To scale this up to an industrial process capable of producing 10,000 tonnes of ammonia per year took two years of incessant labour by a young chemical engineer called Carl Bosch. It was totally just that Bosch should receive a Nobel prize for chemistry, as did Haber, although the latter received the award in 1918 while Bosch was honoured in 1931.

But there is more to chemical engineering than simply scaling up processes from the laboratory to industrial dimensions. Frequently, the actual chemical reactions differ, depending upon the size of the vessels in which they are taking place. Nowhere is this more likely to occur than in chain reactions. If a stable molecule is induced to break into two very reactive parts, which in turn collide with stable molecules to give more unstable radicals, then the possibility exists for a chain reaction with the number of short-lived, unstable molecules growing rapidly. Such a situation can lead to an explosion unless the chains are broken. In a laboratory vessel this may come about by the unstable chain carriers colliding with the vessel walls. In such circumstances, increasing the size of the vessel may permit the reaction to go both further and faster. A reaction which proceeds calmly on the laboratory scale may lead to an explosion in a factory.

To allow for this possibility, the first stage in going from a research synthesis to an industrial process is to build a pilot plant, similar to the laboratory experiment but on a larger scale. Here, any problems to do with stirring the mixtures, adjusting the flow of reactants, or the removal of products can be studied. At the pilot stage, the chemical engineer will also have the opportunity to optimize the efficiency of the chemical process, both in chemical terms and from the point of view of energy costs.

In a working chemical plant it is essential to monitor many parameters, like temperature and pressure, in various parts of the system. Chemical engineers increasingly use computers in their measurement and control systems.

Yet another new dimension is being added to chemical engineering as chemistry becomes more and more involved in biology. There is a new type of biochemical engineer emerging, one who might work on, for example, the process whereby natural gas is converted chemically into methanol using a catalyst. The methanol can then be fed to bacteria which can feed on it to yield a protein-rich product, used as animal feed.

Future of the chemical industry

The chemical industry accounts for less than 5 per cent of the gross national product of a country like Britain; it employs well under 1 per cent of the work-force; but it has a strategic impact on perhaps 40 per cent of the economy. Before the existence of the motor industry, there were no cars. But, by contrast, before the existence of the chemical industry there were clothes, drugs, paints, and insecticides. The chemical industry makes or contributes to the manufacture of all of these. Being a process rather than a product industry its influence is particularly broad and pervasive, and is bound to continue and extend.

The problems which the chemistry industry will face in the future seem to be the rising cost of feedstocks and energy coupled with the public's demand for environmental protection. If one can detect a significant trend in innovation, it must surely be that the industry, like chemistry itself, will work more closely with biology, in part by mimicry.

15

Chemistry in practice

The recurring theme of this short work has been that the basic problem for chemists is to make molecules. In daily laboratory life, various sub-aspects of this essential activity pose their own problems.

Deciding what to synthesize

The chemist begins by knowing or deciding what properties he or she wants in the product being synthesized. The choice of molecule to make may then be obvious. Theoretical considerations may suggest that a particular arrangement of atoms and bonds will provide a product which will have the desired physical or chemical properties. Alternatively, when such theoretical knowledge is lacking, a basic molecular structure can be varied by exchanging one atom or group of atoms for another and then testing the resultant set of new compounds to find the one with the best of the desired properties; when such a strategy, or lack of it, is followed, it is sometimes scathingly referred to as 'molecular roulette'.

An approach is being developed called 'computer-aided molecular design'. This has some parallels with computer-aided design in engineering, particularly since the essential tool is a computer with a graphics-display capability. In areas like pharmaceuticals or agrochemicals, the problem may involve designing a molecule to mimic some known molecule, perhaps a hormone or nerve transmitter. With the aid of computer graphics, it is possible to superimpose representations of molecular structures, using colour and three-dimensional views to facilitate comprehension of similarities and differences. Not only can the framework of atomic nuclei be compared but, in addition, quantum mechanical calculations about the disposition of electron density in and around a molecule can be incorporated in the display. In this way, the hit-and-miss aspect of some synthesis is reduced.

It is the job of the theoretical chemist to perform calculations which give information about the probable shapes and electronic properties of molecules. Almost all such work starts with the famous Schrödinger wave equation that lies at the heart of quantum mechanics. In principle, this equation allows almost any property of a molecule to be calculated using the fundamental ideas of physics. In practice, approximations are made which restrict theoretical calculations to molecules with fifty or less atoms. Solution of the Schrödinger equation starts by giving as input data the types of atoms in the molecule together with their relative positions in space. For this specified arrangement of atoms, the Schrödinger equation yields the energy of the molecule and an associated mathematical function (called the wave function) from which the electron density at any position in or around the molecule may be calculated. Both energy and electron distribution are aids to molecular design.

Choice of the synthetic route

Having decided that it is worth while to make a given molecule, the next problem is deciding how to make it. Generally, the skilful chemist will answer this question using his or her own knowledge and experience, together with a judicious survey of the massive amount of published literature. Here again, however, the role of the computer is growing with the development of what is called retrosynthesis.

In a sense, the problem is not unlike that facing a lawyer who, ideally, would know about all the relevant cases when faced with a legal problem. Whenever there is a vast amount of data, the computer is an obvious device for storing facts and for providing the means of searching through them systematically. Obviously, the computer can never contain details which are not (or were not) in the head of some individual, but it can have facts which are unknown to or have been forgotten by any given individual. Thus the computer ought to be able to suggest alternative synthetic routes, beginning with different starting materials.

A few years ago, retrosynthesis was not especially helpful. But now, after refinement and with more and more facts gradually incorporated into the computer data bases, the most advanced systems are by no means despised by working synthetic chemists.

Determining molecular structure

Having made a single fairly pure compound, the next stage in the chemist's job is to determine the structure of its molecules. This is the most common daily task of chemists. Until recently, the only sure way was to compare the molecules with others of known structure and to perform chemical analysis. Now structure is determined by physical methods. Mass spectrometry and X-ray crystallography have already been briefly mentioned, but by far the most powerful technique is that of nuclear magnetic resonance spectroscopy, as described below.

Some atomic nuclei, such as the hydrogen nucleus (the proton), have the property of spin. A spinning charge acts like a little magnet and hence will be aligned, like a compass needle, by a magnetic field. It requires energy to push an aligned compass needle away from the north–south orientation. The energy of the compass varies from a minimum when pointing north–south to a maximum when it is pointing in the opposite direction, south–north. The same is true for a proton. However, on the microscopic scale, quantum mechanics decrees that any orientation is not possible but only the two extreme alternatives: pointing with the field or against it. The two allowed orientations have different energies.

If the proton is in a strong laboratory magnetic field, then this difference in energy between the two orientations can be determined by shining radio waves on the proton until we find the wave with just the right energy to flip the little nuclear magnet from one allowed orientation to the other. Alternatively, we could shine in waves of one specific radio frequency and vary the magnetic field until the energy difference between the orientations matches the energy of the radio waves—in other words determine the magnetic field strength at which the particular radiowaves were absorbed. If we are dealing with a number of protons and each of them has a different 'electronic environment' in a molecular structure, then there will be a range of magnetic field strengths corresponding to the various protons.

This difference in magnetic field strengths among protons may arise in two ways. All the protons will experience the field of the big external magnet. Within a molecule, however, variations in electron density may shield particular protons from that field to a greater or lesser extent. In addition, one proton may create a magnetic field that influences a nearby proton. The result of these two differentiating factors is a rich source of information which can be readily interpreted in terms of molecular structure. Probably more than any other technique, nuclear magnetic resonance (nmr) has revolutionized the practice of chemistry, making it relatively easy to take a pure substance and, within minutes, to be reasonably sure about the structure of the molecules of which it is composed.

Unsolved problems

Although nmr has proved a wonder-tool, it is not universally appropriate. If molecules are unstable (and consequently short-lived) or if they exist as gases, then nmr may be impotent. The technique requires the high concentrations of molecules found in liquids and solids. As molecules of interest get bigger and bigger, there may also be difficulties in interpretation. Gradually, however, new twists and variations in nmr techniques will make it possible to look at more and more complex molecules—even now, small proteins can be studied.

A major preoccupation of contemporary chemistry is the area of study forming the borderline with biology and the medical sciences. Many of the problems encountered need to be studied with very small quantities of materials and, furthermore, in the natural environments concerned. In consequence, the chemist is constantly trying to develop and adapt his or her techniques both to improve their sensitivity and to cope with complicated mixtures of different types of molecule.

16

Problems and promise

The world as we know it and life in particular grew from a combination of atoms to give molecules. And we have to admit that molecules, some of which have been synthesized and developed by chemists, could equally well end life in its current form—through pollution.

Chemical pollution

Throughout history there have been instances of poisoning and pollution. The affliction of St Anthony's fire and the bizarre happenings with the witches of Salem seem almost certainly to have arisen from poisoning with the alkaloid ergot derived from a fungus which grows on rye. The ergot causes hallucinations and other behavioural abnormalities. Cancer of the liver, common in South-East Asia, can be caused by another fungal toxin, aflatoxin, associated with moulds on peanuts.

Poisoning by heavy metals has rendered some old mining areas barren or unsafe, often due to the presence of impurities in the mined metal. Cadmium deposits in zinc workings, for example, are extremely harmful.

From the study of ecology, it has become clear how poisons which cannot be broken down by the bodies of living creatures are gradually concentrated up the food-chain, reaching catastrophic concentrations at the highest level. Birds of prey accumulate poisons, as do the larger fish and ultimately, if we are the final consumers, man. A striking example is Minimata disease, the widely reported and distressing condition of Japanese fishermen living on tuna, whose flesh contains concentrated mercury.

The situation has worsened since man began to produce new substances not derived directly from nature. Ecologists have discovered how what seemed to be wonderful results of the chemists' work can, in the longer term, sometimes prove deleterious. The pesticide DDT saved the lives of millions of potential malaria sufferers by destroying mosquitoes, but over the years the concentrations of the chemical in birds have begun to produce sterility, and in many countries this pesticide is now banned. Examples of this kind and of the harmful side-effects of drugs, especially thalidomide, have generated a reaction which is more emotional than rational but is none the less powerful. Clearly it is in the public interest for manufacturers to be even more careful about testing every conceivable toxic effect of their products. At the same time, the economic effect is to increase vastly the cost of developing new products and, in consequence, to reduce the amount of innovation. In the world of pharmaceuticals, products which could relieve suffering are not being made available because if by some remote possibility a few individuals were harmed, the manufacturer could face enormous legal and financial penalties.

If the chemist is responsible for creating these problems, then it is the chemist who should solve them. When detergents containing phosphate began to destroy lakes and rivers, chemists had to devise alternatives. In this instance, they designed alternative molecular structures (described technically as 'biodegradable') which could be broken down by bacteria into harmless products and did not, in consequence, produce permanent litter.

Because of the public's right to safety of a high order, all new products have to be tested, creating problems both of a moral and scientific nature. Testing almost invariably implies testing on animals, including such gruesome guarantees as the LD50 test. In this standard and legally required procedure, the quantity of substance lethal to 50 per cent of a population of animals has to be measured. The moral problem arises chiefly with the use of animals to test products for which the need may be slight, such as yet another lipstick; the scientific problem has to do with the meaningfulness of tests on, say, rats when applied to human beings. In the long term, when the biology is better understood, it may be possible to carry out the tests on enzymes in test-tubes, but for the foreseeable future such dilemmas will remain.

Atmospheric pollution

Scientific errors which have led to local pollution and isolated poisoning are distressing. But potentially far more serious are the cases of chemicals which, in very large quantities, may produce total atmospheric or stratospheric pollution, destroying all life on earth.

Already it is evident that burning hydrocarbons can cause serious atmospheric problems. The London smogs of the 1950s were caused by burning coal in domestic fireplaces; this problem has now been regulated. The smog problem in Los Angeles results from the combustion of gasoline allied to a geography and climate which makes it difficult to disperse the smog. Here the chemist's approach has been to try to design catalysts (incorporated in automobile exhausts) which will break down the combustion products of gasoline to carbon dioxide and water. Although superficially this may seem to be the complete answer, it could be nothing more than exchanging one serious problem for another. The carbon dioxide in the atmosphere is increasing owing to the burning of fossil fuels in power-stations and in car engines. At the same time, the world's coverage of greenery, which uses up carbon dioxide in photosynthesis, is decreasing. Unlike the oxygen and nitrogen in the atmosphere, carbon dioxide can absorb heat radiated from the sun or the earth, giving rise to the so-called 'greenhouse effect'. There is fierce debate as to just how serious this problem is, but rapid elevation of the earth's surface temperature could well be cataclysmic.

The particular environmental fear which catches the imagination of the media and the public changes rapidly. The current problem attracting most attention is 'acid rain'. The burning of fossil fuels in power-stations and car engines also produces sulphur dioxide, if sulphur is present in the original fuel. In the atmosphere, the sulphur dioxide gas dissolves in rain-water to produce dilute sulphuric acid, thereby increasing the acidity of the rain. The consequences for the lakes and forests of Scandinavia, Central Europe, and Canada could soon be devastating. Although there is by no means full agreement about either the precise origins of the pollution or the scale of the damage, what is clear is that the problem is a chemical one capable of a chemical solution.

Yet another problem concerns the ozone layer in the earth's atmosphere, which absorbs harmful ultraviolet radiation from the sun and hence protects the earth's surface. The actual amount of ozone is relatively small since the density is very low. Thus it is quite possible for man-made molecules to react with this ozone layer and destroy it—partially, temporarily, or even permanently. Indeed, a few years ago there was an American military proposal to try to do just this in order to disrupt radio communications which use the ozone layer as a reflector. The nullifying of an enemy's radio links could be a powerful ploy in the event of war, but fortunately that experiment was not carried out. More mundane, but equally serious, is the thought that the simple halocarbon molecules used as propellants in aerosols might have the same effect by putting chlorine and fluorine atoms into the stratosphere. This fear is taken sufficiently seriously for the use of propellants to be banned in many states of the USA; meanwhile, thousands of tons of the chemicals are squirted under armpits and up into the heavens every day in the rest of the world, possibly destroying our protective ozone shield.

Warfare

In some senses, the public is always fighting the previous war. There is quite rightly massive concern about the use and deployment of nuclear weapons, but little is heard of chemical weapons, which were the scourge of so many in the First World War. Given the development of chemistry since 1918 and the enormous increase in our comprehension of biology at the molecular level, then here too we should be concerned about the use of science in war, both offensively and defensively.

Yet in chemistry, as in so many other fields, it is war that has provided the stimulus for some of the most far-reaching and long-lasting developments. Before the First World War, the German chemist Haber developed the process of 'fixing nitrogen'—producing ammonia by combining nitrogen in the air with hydrogen. This process has been responsible for solving one of the twentieth century's greatest problems—lack of food for a growing population—by providing nitrogen-containing fertilizers. Haber, however, was equally interested in using his process to manufacture the nitrates vital to the German war effort. Without Haber, the Royal Navy's blockade would have deprived

Germany of explosives by 1916. Haber was also responsible for introducing poison gas into warfare, and after Germany's defeat, he tried personally to pay off her huge indemnity by extracting gold from sea-water; but his idea was based on inaccurate analytical results.

In an almost parallel way in the Second World War, Germany needed to be able to produce rubber substitutes, and again these substitutes were provided by that country's very inventive chemists—taking coke and limestone as raw materials and working through the chemistry of acetylene, which can be made from these cheap sources. Many plastics and antibiotics were developed almost wholly because of the war—even though a balanced view might suggest that the Second World War was more a physicist's war than a chemist's.

Epilogue

Nature is a system which, on a molecular basis, has evolved very slowly. Chemists are now capable of producing very quickly new molecules with predictable structures and properties. We have to be careful not to upset nature's balance through ignorance. In general, chemists have not done so. They have copied nature and collaborated with it, thereby making a widespread and generally beneficial contribution to people's lives.

Chemistry has helped the world's population to be:

- controlled—through contraception
- healthy—through the use of antiseptics and drugs which have all but abolished diseases like tuberculosis
- fed—through the use of fertilizers and insecticides
- clothed—through the use of artificial fibres
- colourful—through the use of modern dyestuffs
- housed—through the construction of buildings incorporating new plastic materials, insulation, and glues
- able to travel—through developments in the chemistry of oil
- able to enjoy leisure—through the use of modern materials in boats, sporting goods, and other leisure equipment.

In short, the chemist's manipulation of molecules has dramatically improved the quality of life for twentieth-century man.

Chemistry is more closely allied to major industry, and hence to politics and economics, than any other major discipline. As such, it is more susceptible to influence than its sister subjects, physics and biology. Because many of the world's largest companies are chemical companies, by no means all the best research is done in universities, and much of the funding for research is private rather than public. Chemistry is thus likely to remain a large and active subject, even if, in the popular view, it is less spectacular than some of the other sciences.

In the future, chemistry seems likely to pursue the broad aims which have become clear over the past few decades, seeking molecular solutions to the problems of society. Two targets seem obvious for such an approach, one inorganic and the other organic. Inorganic chemists, who are increasingly concerned with materials and their properties, are likely to be employed in providing molecular components for electronic devices. The organic chemist will continue to become more closely allied to the biologist, especially in the area of brain research, and as the functioning of the brain becomes better understood in molecular terms, there will be tremendous scope for mimicking and interfering with the natural process. In so far as the brain is a computer, the two branches of chemistry may even coalesce. But whatever the problems of the world may be, since it is a molecular world there ought to be molecular solutions.

Further reading

Abbott, D. (ed.), *The Biographical Dictionary of Scientists-Chemists*, Blond Educational, London (1983).

Asimov, I., *A Short History of Chemistry*, Doubleday, London (1972).

Dawkins, R., *The Selfish Gene*, Oxford University Press (1976).

Dickerson, R. E., and Geiss, I., *Chemistry, Matter, and the Universe*, Benjamin, Menlo Park (1976).

Dickerson, R. E., Gray, H. B., and Haight, G. P., *Chemical Principles* (3rd edn.), Benjamin/Cummings, Menlo Park (1979).

Gamov, G., *Mr. Tompkins in Paperback*, Cambridge University Press (1965).

Gardner, M., *The Ambidextrous Universe*, Allen Lane, London (1967).

Gasser, R. P. H., and Richards, W. G., *Entropy and Energy Levels*, Oxford University Press (1974).

Hill, G. C., and Holman, J. S., *Chemistry in Context* (2nd edn.), Thomas Nelson and Sons, Walton-on-Thames (1983).

Pilar, F. L., *Chemistry, the Universal Science*, Addison-Wesley, Reading, Mass. (1978).

Rose, S., *The Chemistry of Life* (2nd edn.), Penguin, Harmondsworth (1979).

Rossotti, H., *Introducing Chemistry*, Penguin, Harmondsworth (1984).

Sharp, D. W. A. (ed.), *The Penguin Dictionary of Chemistry*, Penguin, Harmondsworth (1983).

Watson, J. D., *The Double Helix*, Norton, London (1980).

Index